# BASIC matrix methods

# BASIC matrix methods

Including applications in approximation and data fitting

J C Mason, MA, DPhil, FIMA

Principal Lecturer
Department of Mathematics and Ballistics
Royal Military College of Science
Shrivenham
Swindon
Wilts

**Butterworths**
London . Boston . Durban . Singapore . Sydney . Toronto . Wellington

First published 1984
  Reprinted 1987

© Butterworth & Co (Publishers) Ltd, 1984

---

**British Library Cataloguing in Publication Data**
Mason, J. C.
  BASIC matrix methods.
  1. Matrices—Data processing      2. Engineering
  mathematics—Computer programs      3. Basic
  (Computer program language)
  I. Title
  512.9′34′0285424      TA347.D4

  ISBN 0–408–01390–7

---

**Library of Congress Cataloging in Publication Data**

Mason, J. C.
  BASIC matrix methods.

    1. Matrices—Computer programs.      2. Basic (Computer
  program language)      I. Title.      II. Title: B.A.S.I.C. matrix
  methods.
  QA188.M38 1984      512.9′434′0285425      83–15059
  ISBN  0–408–01390–7

---

Filmset in Monophoto Times by Northumberland Press Ltd, Gateshead
Printed in Great Britain by The Camelot Press Ltd, Southampton

# Preface

Matrices are an invaluable tool in the computer solution of a wide variety of mathematical problems in science and engineering. In particular, matrix methods are needed for the solution of the systems of linear algebraic equations (i.e. simultaneous equations) which occur in many models of physical processes, and two significant areas of general application are in approximation and data fitting and in the numerical solution of differential equations.

The primary aims of the present book are to introduce the reader to matrices and their uses and to develop BASIC programs to implement various matrix methods. A substantial part of the book is devoted to the solution of linear algebraic systems by both direct and iteration methods, and eigenvalue problems are also covered. In addition the book includes a discussion of matrix methods in approximation and data fitting.

The book serves as a natural sequel to *BASIC Numerical Mathematics* while remaining essentially independent of it. Like its predecessor, it is geared to the requirements of undergraduate engineers in the overall spirit of the books in the Butterworth BASIC series. However, it should also meet the needs of a wide range of engineers, mathematicians, and scientists who wish to solve practical problems involving matrices on a computer. It should be particularly useful as a text for numerical mathematics courses, engineering problems courses, and computer applications courses.

There are 18 well documented BASIC programs in the book for executing many of the fundamental matrix methods required by engineers. The book also includes a reasonably thorough discussion of the mathematical principles behind these methods and a good selection of problems of both mathematical and practical types.

The BASIC language, which is adopted in this book, is widely available on both mainframe computer systems and minicomputers. It is very straight-forward to write and modify computer programs in BASIC, and the reader should not encounter any special difficulty in making use of the programs in the book. All programs have been written in an 'interactive' fashion, so as to make them easy to use for demonstration and teaching purposes. Following the instruction RUN the user is asked for all necessary data, and the input and output of all information takes place at the keyboard. Readers who need batch-mode programs (which include all their own data) may easily modify the programs in the book by using READ and DATA statements in place of INPUT statements, and examples of suitable codes may be found in Chapter 4.

One of the present drawbacks of BASIC lies in its poor subroutine facilities. This shows up in Chapter 7, where BASIC matrix routines are used to solve linear algebraic systems, since the programs developed in Chapters 5 and 6 cannot be used as subprograms. The codes of the latter programs can, however, be used if they are typed in in full, and this is what we would recommend to users who do not have BASIC matrix routine facilities.

Readers who wish to look up programs in the book are referred to the contents list for page references. The programs themselves are often placed within the body of the text rather than before or after the corresponding theory. They are typically preceded by algorithms (with the same numbering) which list the logical steps, and followed by Sample Runs which demonstrate the programs in action and by Program Notes which give necessary additional information and advice.

As in *BASIC Numerical Mathematics*, we have aimed to strike an appropriate balance between numerical methods and numerical analysis, both of which are important to engineers. We have also tried to whet the reader's appetite in all the problem areas, although this has sometimes resulted in our using simple but imperfect methods (as in the use of BASIC matrix routines in Chapter 7 to solve the 'normal equations' of the least squares method). However, we have tried to draw attention to the limitations of each method, while giving references to more advanced techniques.

A large number of problems are posed at the ends of chapters, and these include examples of a number of engineering applications. Many of the problems are designed to help the reader understand and use the theory and programs in the book. However, several problems aim to increase the reader's mathematical knowledge and put new prospectives on the theory, and several opportunities are given to extend and develop the given programs.

As far as the actual contents of the book are concerned, these consist essentially of four short introductory or elementary chapters, followed by three substantial chapters covering the major topics of the book. Chapters 1 to 3 give introductions to the three key aspects of the book: the BASIC computer language, numerical mathematics, and matrices. Chapter 4 then develops a number of BASIC routines for elementary matrix calculations, as well as providing practice in BASIC programming.

Chapter 5 covers the vital subject of the direct solution of linear algebraic systems. It includes a brief demonstration of the use of BASIC matrix routines, followed by a detailed description of the Gauss elimination method. Considerable care is taken in this chapter to provide programs which not only produce 'answers' but also tell the user whether or not these answers are likely to mean anything. Although the direct Gauss elimination method is closely related to the 'school method' for solving simultaneous equations, some readers may prefer to master the simpler indirect methods first. Such methods are covered in Chapter 6, not only for the solution of linear algebraic systems but also for the treatment of eigenvalue problems.

Chapter 7 extends the familiar idea of fitting a straight line through a set of data, and considers the problem of fitting polynomial, spline, and general curves to mathematical functions as well as to sets of data. The emphasis here is on collocation and least squares methods, based on the solution of appropriate linear algebraic systems by matrix methods.

We are again grateful to a number of colleagues at the Royal Military College of Science in Shrivenham for helping with this book in various ways. We should especially like to extend our thanks to Mrs Jan Price for preparing the manuscript with great patience and skill, and to Mr D. C. Stocks, Dr M. J. Iremonger, and Mr P. D. Smith for reading and criticising various parts of it.

J.C.M.

In memory of my parents

# Contents

PROGRAMS

# Chapter 1

# Introduction to BASIC

## 1.1 Computer programs and programming languages

A computer program is a set of instructions which a computer is able to interpret and execute. These instructions are designed to perform a particular task, and in our case this task is to determine the numerical solution of a mathematical problem using matrices. In order that the instructions may be recognised, they must be written in a standard programming language (such as BASIC, FORTRAN, ALGOL, PASCAL, COBOL, etc.) for which an 'interpreter' (or 'compiler') is available on the computer. The interpreter transforms each instruction in the programming language into a set of fundamental instructions in a 'machine language', designed to be instantly recognisable to the computer. The programming language is designed to be convenient and practicable for the user and BASIC is one such programming language.

## 1.2 The BASIC approach

All of the programs in this book are written in BASIC. The name BASIC is an acronym for Beginner's All-purpose Symbolic Instruction Code, and it was developed at Dartmouth College USA as a general purpose computer language. The main advantages of BASIC are that it is easy to learn, convenient to use, and particularly well suited to 'conversational' programming in which the user interacts with the computer throughout the running of the program.

The simple version of BASIC used in this book has a number of disadvantages, and these mainly concern its lack of structure in comparison with languages like FORTRAN or PASCAL. For example, it is not usual in BASIC to distinguish between integers and other numbers, to have variables of double length (for more accurate calculations), or to use one program as a subroutine or subprogram for another program. Moreover, BASIC has a particular disadvantage in numerical analysis, which relates to its apparently commendable feature of rounding to integer values any numbers that are very close to integers. This makes it difficult to test the conditioning of any problem that has integer data, and inadvisable to use integer data as test data in gauging the rounding error in any program. However, these disadvantages are not too important a consideration for elementary programs such as those given in the following chapters.

Although the book does not give every detail of the grammar of BASIC, a description of the main features of BASIC is given below. For a more detailed approach the reader is referred to References 1 to 3.

## 1.3 The elements of BASIC

### 1.3.1 *Program structure and sequencing*

A BASIC program is a sequence of statements which define a procedure for the computer to follow. As it follows this procedure, the computer allocates values to each of the variables encountered and changes them where instructed. Statements used in the program are of a number of types, which will be discussed in more detail in following sections. They include REM statements (for making program notes), DIM statements (for allocating subscripted variables), INPUT or READ statements (for defining data), assignment statements (for doing mathematics), conditional statements (for controlling the action of the program) and PRINT statements (for printing out results).

Every statement must be preceded by a line number. On running the program, all statements are executed in the sequence that corresponds to these line numbers. For example, the program

$$100 \ X = 1 \qquad \text{is executed as} \ 100 \, X = 1$$
$$400 \ \text{GO TO } 200 \qquad\qquad\qquad 200 \ X = X + 1$$
$$300 \ \text{PRINT X} \qquad\qquad\qquad 300 \ \text{PRINT X}$$
$$200 \ X = X + 1 \qquad\qquad\qquad 400 \ \text{GO TO } 200$$

The use of numbering greatly simplifies correcting and editing (see Section 1.5).

### 1.3.2    *Mathematical expressions*

In mathematics it is necessary to evaluate expressions which involve numerical constants, variables (e.g. $X$), and functions (e.g. SIN). All constants are treated identically in BASIC, whether they are integer (e.g. 36) or real (e.g. 36.1). They may be entered in either fixed point form (e.g. 36.1) or floating point form (e.g. 0.361 E2), although the computer prints out numbers in fixed point form unless they are small or large. The constant $\pi$ is often available by typing PI or the $\pi$ key.

Variables, which fulfil the role of letters in algebra, may be named by any one of the letters A to Z, or by any letter followed by a digit (e.g. A3, P7, etc). Each variable is allocated a location in the computer store, and it takes the numerical value recorded in that location. This numerical value is substituted for the corresponding variable whenever that variable occurs in an expression, and so it is important to ensure that the correct value is given to a variable initially.

The function square root may be evaluated via the built-in computer function SQR, $\sqrt{x}$ being replaced by SQR(X). The argument in brackets (X) may be any number, variable, or mathematical expression. Other built-in functions include SIN(X), COS(X), LOG(X), EXP(X), ABS(X), and INT(X) which represent, respectively, $\sin x$, $\cos x$, $\ln x$ (i.e. $\log_e x$), $e^x$, $|x|$, and the integer part of $x$. For trigonometric functions (SIN, etc) the argument is assumed to be measured in radians.

Mathematical expressions are formed from constants, variables and functions by inserting arithmetic operations such as plus, times, etc. These operations have a hierarchy, which determines the order in which they are performed by the computer, and it is as follows:

to the power of ($^\wedge$)
multiply (*) and divide (/)
add (+) and subtract (−).

If two or more operations have the same hierarchy, then the computer works from left to right. Brackets always take precedence and should be used to provide clarity and avoid ambiguity. The first left bracket is paired with the last right bracket, and so on. Hence

$$\frac{a + b}{3c}$$

becomes either

$$(A + B)/(3*C) \text{ or } (A + B)/3/C.$$

Some examples of correct and incorrect BASIC expressions are as follows:

| Mathematical Expression | Correct BASIC | Incorrect BASIC |
|---|---|---|
| $x \sin x$ | X * SIN(X) | X SIN(X) |
| $\dfrac{1 - r^n}{1 - r}$ | (1 − R$^\wedge$N)/(1 − R) | 1 − R$^\wedge$N/1 − R |

| Mathematical Expression | Correct BASIC | Incorrect BASIC |
|---|---|---|
| $\ln(1 + \sqrt{x})$ | LOG (1 + SQR(X)) | LOG (1 + SQR(X) |
| $\left\lvert \dfrac{1 + \sin x}{x} \right\rvert$ | ABS((1 + SIN(X))/X) | ABS (1 + SIN(X))/X |

### 1.3.3 *Assignment statements*

An assignment statement takes the form

line number [LET] variable = mathematical expression

The word LET here is usually optional, and will be omitted throughout this book. Square brackets are used in this chapter to indicate optional items. For example a root of a quadratic equation

$$x_1 = (-b + \sqrt{b^2 - 4ac})/(2a)$$

may be obtained by a statement such as

100 X1 = (−B + SQR(B^2 − 4*A*C))/(2*A)

It is important, however, to realise that an assignment statement is not itself an equation. It is an instruction to give the variable on the left-hand side the numerical value of the expression on the right-hand side. Thus we may have a statement such as

50 X = X + 1

which increases by 1 the value of X.

There is no mathematical statement in common usage which is precisely equivalent to the assignment statement

X = Y

However, in this book we shall use the symbol ':=' to denote 'becomes', so that the precisely equivalent mathematical statement is

$x := y$

The symbol ':=' is used in place of '=' for assignment statements in the ALGOL language.

### 1.3.4 *Input*

In *conversational programming* the user specifies values of variables by INPUT statements at 'run-time'. The statement has form

line number INPUT variable 1 [, variable 2, ...]

e.g.

    20 INPUT A, B, C

When the program is run the computer prints ? on reaching this statement, and waits for the user to type values for the variables, e.g. ? 5, 10, 15 which are interpreted as A = 5, B = 10, C = 15 in the above example.

    An alternative form of data input is used if there are many data, or if the data are unlikely ever to be changed, or if the user does not want to converse with the computer. In this case a statement of the form

    line number READ variable 1[, variable 2, ...]

e.g.

    20 READ A, B, C                  (1.1)

is used in conjunction with a statement (or number of statements) of form

    line number DATA number 1 [, number 2, ...]

e.g. either 21 DATA 5, 10, 15            (1.2)
or       21 DATA 5   ⎫
        22 DATA 10  ⎬     (1.3)
        23 DATA 15  ⎭

On executing a READ statement, values are assigned to variables from the DATA statements in the order in which the latter occur in the program. If (1.1) is followed by either (1.2) or (1.3), then A, B and C are allocated values 5, 10, and 15.

### 1.3.5 *Output*

The output of data (for checking purposes) and the results of calculations etc. is done by a statement of form

    line number PRINT list

where the list may contain variables or expressions

e.g.

    200 PRINT A, B, C, A*B/C

or text enclosed in quotes

e.g.

    10 PRINT "VALUES OF A, B, C:";

or a mixture of both

e.g.

300 PRINT "X = "; X, "Y = "; Y

The items in the list are separated by commas or semi-colons. Commas ensure a tabulation in columns about 14 spaces wide, while a semi-colon suppresses such spacing. If a semi-colon is placed at the end of a list, it suppresses the line feed. If the list is left blank then a blank line is printed, and this is a useful way of spacing out results.

Note the necessity to use PRINT statements to copy out all numbers which are input by INPUT or READ/DATA statements, so that there is a true record of them. PRINT statements should also precede INPUT statements for explanatory purposes, since ? on its own is not informative. For example the pair of statements

10 PRINT "WHAT IS X";
20 INPUT X

lead to the computer output

WHAT IS X?

in reply to which the value of X is typed in.

### 1.3.6 Conditional statements

It is often necessary to take a course of action if, and only if, some condition is fulfilled. This is done with a statement of form

line number IF expression 1 $^{\text{conditional}}_{\text{operator}}$ expression 2 THEN line number

where the possible 'conditional operators' are

$=$     equals
$<>$   not equal to
$<$     less than
$<=$   less than or equal to
$>$     greater than
$>=$   greater than or equal to

For example a program could contain the following statements if it is to stop when a zero value of N is input.

20 INPUT N
30 IF N < > 0 THEN 50          (1.4)
40 STOP
50 ...

Note the statement STOP which ends the physical run of a program. This is not to be confused with the statement END which is the (optional) last statement occurring in the program listing.

### 1.3.7 *Loops*

There are several ways in which a program may be made to repeat some of its procedure, and the simplest is to use the statement

line number GO TO line number

For example, if the statement

60 GO TO 20

is added to the instructions (1.4), then the program will execute statement 50 for a sequence of input values of N until a zero is input.

The most common way of doing loops is to start with a 'FOR statement'

line number FOR variable = expression 1 TO expression 2 [STEP expression 3]

where the STEP is assumed to be unity if it is omitted, and end the loop with

line number NEXT variable

The same variable is used in both FOR and NEXT statements, and its value should not be changed in the intervening statements.

A loop is used if, for example, N sets of data X, Y have to be read and a mathematical expression such as sin $(X + Y)$ calculated in each case.

e.g.

```
10 READ N
20 PRINT "X", "Y", "SIN (X + Y)"
30 FOR I = 1 TO N
40 READ X, Y
50 PRINT X, Y, SIN (X + Y)
60 NEXT I
```

Loops may also be used to calculate sums and products of a list of expressions, and this is discussed in Chapter 4 in connection with matrix additions and multiplications.

### 1.3.8 *Subscripted variables*

It is frequently desirable in mathematics to use a variable with a subscript, such as $x_i$, so that many cases can be covered by a simple formula. For example, we might write

$$x_i = i^2 \quad (i = 1, 2, 3, \ldots, 10) \tag{1.5}$$

to specify that the $x_i$ are the squares of the integers up to 10 ($x_1 = 1$, $x_2 = 4, x_3 = 9, \ldots, x_{10} = 100$). In a BASIC program $x_i$ is represented by X(I), the subscript being placed in brackets, and a specific numerical value must be assigned to I in the program, perhaps by a FOR loop. For example (1.5) may be calculated from

```
10  FOR I = 1 TO 10
20  X(I) = I^2
30  NEXT I
```

It is also permitted for a variable to have two or more subscripts attached to it, such as A(I, J).

Since a subscripted variable has more than one value associated with it (while a non-subscripted variable has just one), it is necessary to provide computer storage space for as many values as might be needed. This is done by a 'dimension statement' of the form

line number DIM variable 1 (integer 1) [, variable 2 (integer 2), ...]

e.g.

10 DIM X(50), Y(50), A(10, 10)

which allows up to 51 values X(0), ..., X(50), up to 51 values Y(0), ..., Y(50), and up to 121 values A(0, 0), ..., A(10, 10). The DIM statement must occur before the first use of the subscripted variables.

In some computers it is possible to do 'dynamic dimensioning' with a statement like

20 DIM X(N), Y(N)

provided that N has been previously defined, and this form is useful for avoiding wasted storage space.

### 1.3.9 *Matrix variables*

There are no 'matrix variables' as such in BASIC. However, subscripted variables with two subscripts fit the bill precisely, and so we shall refer to these as matrix variables whenever they are used to represent matrices. In defining a matrix **A**, it is necessary to specify the number $m$ of its rows, the number $n$ of its columns, and the $mn$

entries $a_{ij}$ $(i = 1, \ldots, m; j = 1, \ldots, n)$. On the computer, each entry $a_{ij}$ is stored in the corresponding subscripted variable A(I, J), and the values M and N of $m$ and $n$ are normally fixed in the corresponding DIM statement.

To illustrate the assignment of a matrix, the following simple code defines **A** to be the unit 3 by 3 matrix $I_3$ (with ones in diagonal positions $a_{11}$, $a_{22}$, $a_{33}$ and zeros elsewhere).

```
10 DIM A(3, 3)
20 FOR I = 1 TO 3
30 FOR J = 1 TO 3
40 A(I, J) = 0
50 NEXT J
60 A(I, I) = 1
70 NEXT I
```

### 1.3.10 *Subroutines*

Sometimes it is necessary to use a certain sequence of statements more than once in a program, and, in order to avoid repeating these statements, it is sensible to use a subroutine for this sequence. The program then contains statements such as

line number GOSUB line number

which causes the program to transfer control to a set of instructions (the subroutine) which starts at the second line number. The subroutine must end with an instruction of the form

line number RETURN

and the program then returns control from the subroutine to the statement immediately after the GOSUB call.

### 1.3.11 *Other statements and facilities*

(a) *REM statements* are used for explanatory comments or headings in the program listing, and have the form

line number REM comment

e.g.

10 REM—THIS PROGRAM SOLVES Y′ = F(X, Y), Y(0) = 1

Such statements are ignored when the program is run. In some computers comments may be included on the same line as other statements.

(b) *String variables* enable the use of non-numeric data (e.g. words) and may be used, for example, for reading a combination of numbers and words. They will not be used in this book.

(c) *Multiple branching* can be done with statements of the form

line number ON expression THEN line number 1
[, line number 2, ...,]

or

line number ON expression GOSUB line number 1
[, line number 2, ...,]

The program transfers to the line number 1, line number 2, ... according as the integer value of the expression is 1, 2, .....

(d) *Function definition statements* are important in mathematics since they enable us to define our own functions (in addition to built-in functions such as SIN(X)). New functions may have any of the names FNA, FNB, ..., FNZ and are created by a statement such as

$$10 \text{ DEF FNA}(X) = .5*(\text{EXP}(X) + \text{EXP}(-X)) \tag{1.6}$$

which forms the function cosh(x). Any defined function, e.g. FNA(X) above, is simply used in the main body of the program in a statement such as

$$100 \text{ Y} = \text{FNA}(1)$$

which sets Y equal to cosh(1) if FNA is defined by (1.6).

Many computers allow functions of several variables. Thus the function $f(x, y) = x^2 + y^2$ might be named in the program as FNF(X, Y) and defined by the statement

$$20 \text{ DEF FNF}(X, Y) = X^2 + Y^2$$

## 1.4 Matrix routines

On many computers built-in matrix routines are available in BASIC. However, since these routines requires considerable storage, they are not always available on microcomputers. Nevertheless they are a useful tool in matrix methods, as we shall see in subsequent chapters.

Great care needs to be taken in the correct dimensioning of matrix variables via DIM statements. It is also recommended that the manual appropriate to a particular computer should be studied, since details vary between computers. For example some computers require a vector X with 10 components to be dimensioned as X(10, 1) with 2 subscripts (i.e. 10 rows and 1 column) whenever it is to be pre-multiplied by a square matrix, say A(10, 10), while other computers permit just one subscript X(10) to be used.

The main BASIC matrix instructions are summarised in Table 1.1. In the list of assignment statements, a different matrix variable A is used on the left-hand side from the matrix variables B and C on the right-hand side. However, it is in fact permissible to use the same matrix variables on both sides, as in instructions such as MAT A = A + B and MAT A = INV(A). (In these respective instructions, **A** and **B** are added and the resulting matrix is stored in **A**, and **A** is replaced by its inverse $\mathbf{A}^{-1}$.)

**Table 1.1** *Matrix statements*

| Mathematics | BASIC | |
| --- | --- | --- |
| **A** = **B** | MAT A = B | |
| **A** = **B** + **C** | MAT A = B + C | |
| **A** = **B** − **C** | MAT A = B − C | |
| **A** = K **B** (K scalar) | MAT A = (K)*B | |
| **A** = **B** **C** | MAT A = B*C | |
| **A** = **O** (all zeros) | MAT A = ZER | |
| Read **A** | MAT READ A | ⎫ Elements are listed |
| Input **A** | MAT INPUT A | ⎬ individually by rows |
| Print **A** | MAT PRINT A | ⎭ |
| **A** = m by n matrix of ones | MAT A = CON(M, N) | |
| **A** = **B**ᵀ | MAT A = TRN(B) | |
| **A** = **B**⁻¹ | MAT A = INV(B) | |
| **A** = **I** | MAT A = IDN | |

Note that in matrix routines all subscripts are taken to be numbered from 1 upwards (to correspond to row or column numbers). Thus the pair of statements

    10 DIM A(10, 10)
    20 MAT INPUT A

will input 100 numbers A(1, 1), ..., A(1, 10), A(2, 1), ..., A(10, 10).

## 1.5 Checking and editing programs

If a program has grammatical (syntax) errors in it, then the computer will normally give a clear indication of them. Care needs to be taken, however, to distinguish correctly between the letter 'oh' and the number 'zero' and between 'i' and 'one'. Also mystifying errors may occur if a variable is used for several purposes in the same program.

It is not of course sufficient for a program to be grammatically correct. It must give correct results, and should therefore be tested by using simple test data that give a known solution or by performing hand calculations with simple test data. It is also desirable to ensure that the program cannot go out of control for foolish choices of data, such as a negative number for the number of equations in a problem, and that it is able to cover as wide a range of potential data as possible. It is quite difficult to make programs completely 'userproof', and they become long in doing so. The programs in this book have been kept as short as possible, while providing adequate explanation, and so they are not always 'userproof'.

## 1.6 Different computers, and variants of BASIC

The algorithms and examples in the book are programmed in a simple version of BASIC that should work on most computers, even those with small storage capacity. Only single line statements have been used, although many computers allow a number of statements on each line with a separator such as \ or :. Multiple assignments are also sometimes allowed so that, for example, the variables A, B, C, D are each set to zero by the statement

70 A = B = C = D = 0

There is one important area in which computers vary, and this is particularly relevant to microcomputers with visual display units (VDUs). This concerns the number of available columns across each line and the number of rows that are visible on the screen. Simple modifications may be necessary to fit the output of some of the programs in this book on a particular microcomputer. For this purpose 'TAB' printing is a useful facility.

Various enhancements of BASIC have appeared over the years and these are implemented on many computer systems. Indeed the programs in this book could be re-written to incorporate some of these advanced features. For example, the availability of long variable names (e.g. ROOT instead of say X) can make it easier to write unambiguous programs, although it may be argued that one-character symbols provide a closer link with algebra. Other advanced facilities include more powerful looping with conditional statements, and independent subroutines.

Independent subroutines are particularly valuable in numerical mathematics where one often wishes to call upon a useful program, such as one which solves a set of linear algebraic equations (see Chapter 5) in another program. Indeed, in Chapter 7 below we have had to use BASIC matrix routines to solve a set of linear equations, rather than the programs specially developed in Chapters 5 and 6, because of the lack of suitable subroutine facilities in standard BASIC. If and when independent subroutine facilities become available, the programs in Chapter 7 could be suitably modified.

The use of independent subroutines leads to the disciplines of 'structured programming', in which programs are divided into a number of smaller independent subprograms which are called upon in turn from the main 'driver' program. This greatly eases the task of debugging, since it is often possible to base a new program on a number of old well-tested subprograms.

## 1.7 Summary of program contents

In writing a BASIC program, the order in which we go about listing instructions is roughly as follows:

   (i) REM statements at start (and throughout program) for explanation
  (ii) DIM statements for subscripted variables
 (iii) INPUT or READ/DATA statements for data
  (iv) *Main program*, with calculations, etc.
   (v) PRINT statements for results
  (vi) STOP statement to end calculation (unnecessary at the end of a program)
 (vii) DEF statements for function definitions
(viii) Subroutines with RETURN statements
  (ix) END statement (optional)

However, although the above order is our preferred one, some implementations of BASIC still require DEF statements (or subroutines) to *precede* any statements in which defined functions (or subroutines) are used. On such BASIC systems it will be *necessary to change the line numbers* of all DEF statements in programs in this book. For example, the line numbers 480 and 490 of the DEF statements in Program 7.1 will need to be changed to line numbers such as 51 and 52. (The same remark applies to programs in our earlier book *BASIC Numerical Mathematics*.)

## 1.8 References

1. Kemeny, J.G. and Kurtz, T.E., *BASIC programming*, John Wiley (1968).
2. Monro, D.M., *Interactive Computing with BASIC*, Edward Arnold (1974).
3. Alcock, D., *Illustrating BASIC*, Cambridge University Press (1977).

Chapter 2

# Introduction to numerical mathematics

**ESSENTIAL THEORY**

Matrix methods are discussed in this book primarily in the context of numerical mathematics, and so it is appropriate to give the reader a brief introduction to the fundamentals of the latter. For a more detailed treatment and also a selection of problems, the reader is referred to Reference 1, especially if he or she is uncertain about such elementary ideas as decimal and binary numbers, and absolute and relative errors.

## 2.1 The task

Given a mathematical problem, the task in numerical mathematics is to obtain the actual numerical values of the solution. This task has two main aspects:

    (i) to *find a feasible method* of determining the numerical solution,
    (ii) to *analyse* both the method and the computed solution.

Broadly speaking, aspect (i) is what is meant by the subject of *numerical methods* and aspect (ii) is what is meant by the subject of *numerical analysis*. However, since (i) has little meaning without (ii), we should always give due consideration to both aspects.

### 2.1.1 *Numerical methods*

Numerical mathematics, which is essentially a branch of mathematics, differs from much of 'traditional mathematics' in putting its main emphasis on the requirements of the actual computation. In particular, a feasible numerical method must have the following properties:

    (i) it must be *efficient*,
    (ii) it must involve a *finite* number of operations.

    In traditional mathematics, methods are sometimes described, typically for conceptual or academic reasons, which are very in-

14

efficient from the point of view of practical computation. A prime example is *Cramer's rule* for the solution of a system of linear algebraic equations, based on the recursive calculation of determinants. A discussion of this method was given in Reference 1, but it is repeated here because of its immediate relevance. Cramer's rule solves such a system (see (3.18) below) in the form

$$
\begin{array}{cc}
x_1 & x_2 \\
\begin{vmatrix} b_1 & a_{12} & a_{13} & \cdots & a_{1n} \\ b_2 & a_{22} & a_{23} & \cdots & a_{2n} \\ \vdots & & & & \vdots \\ b_n & a_{n2} & a_{n3} & \cdots & a_{nn} \end{vmatrix} = & \begin{vmatrix} a_{11} & b_1 & a_{13} & \cdots & a_{1n} \\ a_{21} & b_2 & a_{23} & \cdots & a_{2n} \\ \vdots & & & & \vdots \\ a_{n1} & b_n & a_{n3} & \cdots & a_{nn} \end{vmatrix} = \cdots
\end{array}
$$

$$
\begin{array}{cc}
x_n & 1 \\
= \begin{vmatrix} a_{11} & a_{12} & \cdots & a_{1,n-1} & b_1 \\ a_{21} & a_{22} & \cdots & a_{2,n-1} & b_2 \\ \vdots & & & & \vdots \\ a_{n1} & a_{n2} & \cdots & a_{n,n-1} & b_n \end{vmatrix} = & \begin{vmatrix} a_{11} & a_{12} & \cdots & a_{1n} \\ a_{21} & a_{22} & \cdots & a_{2n} \\ \vdots & & & \vdots \\ a_{n1} & a_{n2} & \cdots & a_{nn} \end{vmatrix}
\end{array} \quad (2.1)
$$

Each of the unknowns $x_1, x_2, \ldots, x_n$ is determined *explicitly* as a ratio of a pair of determinants, but unfortunately the number of multiplications (and additions) involved in calculating the relevant determinants is very great. Specifically each of these $n + 1$ determinants of order $n$ must be expanded (see Section 3.6) in terms of $n$ determinants of order $n - 1$, which must each be expanded in terms of $(n - 1)$ determinants of order $n - 2$, and so on. Thus the total number of multiplications involved is $(n + 1)!$, which exceeds thirty million for $n = 10$. In contrast the Gauss elimination method, which will be discussed in Chapter 5, involves only about $n^3/3$ multiplications, or around three hundred multiplications for $n = 10$. Thus Cramer's rule is some 100,000 times as inefficient as Gauss elimination for the solution of 10 equations. (We shall see in Chapter 5, that the determinant of any $n \times n$ matrix might in fact be calculated in $n^3/3$ multiplications by using the Gauss elimination method. However, it would still be absurd to implement the Gauss elimination method $n + 1$ times on order to perform Cramer's rule, since the Problem (3.18) may be solved by *one* application of Gauss elimination.)

It is also common in traditional mathematics to propose methods based on infinite processes for the solution of a problem. For example the solution $y(x)$ of the differential equation

$$\frac{dy}{dx} = 1 - 2xy, \quad y(0) = 1 \tag{2.2}$$

may be obtained for all $x$ in $[0, \infty)$ from the explicit formula

$$y(x) = \exp(-x^2) \int_0^x \exp(t^2) dt \tag{2.3}$$

Here exp $(x)$ denotes the exponential function $e^x$.

However, (2.3) involves the indefinite integral of the function $\exp(t^2)$ which cannot be determined exactly by a simple analytical method. Thus we have replaced the problem (2.2) by one (namely (2.3)) which is not necessarily easier to handle. Indeed the solutions of both (2.2) and (2.3) inevitably involve infinite processes. In implementing a numerical method we must thus obtain an *approximate* solution to one or other of these problems by replacing the relevant infinite process by a finite one.

### 2.1.2 *Numerical analysis*

Probably the most important part in the numerical analysis of a problem and method is the study of the *error* between the computed approximate solution and the true solution. This error is related both to the efficacy and efficiency of the method and to the meaningfulness of the problem. Moreover it is particularly desirable to be able to perform an *error analysis*, namely a determination of quantitative estimates or bounds for the error in the solution. To do this we must understand and analyse the various sources of error in a calculation. Sometimes such an error analysis may be performed *before* the calculation of the solution, and in that case it is termed an *a priori* error analysis. It may also be based on the actual results produced, and then it is termed an *a posteriori* error analysis.

### 2.2 Sources of error

A principal source of error in numerical mathematics is of course human error. This may be the consequence, for example, of misreading or misinterpreting a method, of using a method which is actually wrong, or of incorrectly coding the computer implementation of a method. Clearly this type of error should be eliminated before any others are considered.

The three other main sources of error are (i) *data errors*, (ii) *truncation errors*, and (iii) *rounding errors*. These relate in turn to three key ideas in numerical analysis: (i) *conditioning*, (ii) *convergence*, and (iii) *stability*.

First let us consider data errors. A problem is often not defined exactly, and in particular the data are sometimes approximate numbers. Such errors may obviously affect the solution. Indeed if a small change in the data leads to a small change in the solution then we say that the problem is *well-conditioned* and if it leads to a large change in the solution then we say that the problem is *ill-conditioned*. It is not uncommon for systems of linear algebraic equations to be ill-conditioned, and we shall see examples of this in Chapter 5. Clearly data errors can have serious consequences in an ill-conditioned problem, and so it is important to be honest in recognising ill-conditioning. Note also that, since it is a feature of the problem, ill-conditioning cannot be eradicated by changing method, unless this results in the problem being restated in a new form.

Truncation errors, on the other hand, are a feature of the method rather than the problem, and more specifically they concern its mathematical properties. By a truncation error we mean an error which is committed in replacing an infinite process (which cannot be computed) by a finite process (which can be computed) with say $n$ steps. For example if we calculate $y = e$, the base of natural logarithms, by truncating the infinite expansion

$$y = e = 1 + \frac{1}{1!} + \frac{1}{2!} + \frac{1}{3!} + \ldots + \frac{1}{n!} + \ldots \qquad (2.4)$$

into the finite approximation $y_n$ given by

$$y_n = 1 + \frac{1}{1!} + \frac{1}{2!} + \ldots + \frac{1}{n!} \qquad (2.5)$$

then we commit a truncation error of $y - y_n$ given by

$$y - y_n = \frac{1}{(n + 1)!} + \frac{1}{(n + 2)!} + \ldots \qquad (2.6)$$

Here $y - y_n$ approaches zero as $n$ approaches infinity, as a consequence of the convergence of the series (2.4). In general, we therefore say that a method is *convergent* if the truncation error approaches zero as the number $n$ of steps approaches infinity. The ideas of convergence and truncation error will be important in Chapter 6 when we consider iterative (i.e. repetitive) methods for solving linear algebraic equations and eigenvalue problems, but they will be irrelevant in Chapter 5 where direct methods are used.

Finally, rounding error is the error which occurs in the computer implementation of a method as an inevitable consequence of the use, throughout the calculation, of inexact computer arithmetic. For all arithmetic operations such as multiplication and addition are performed inexactly in general, so as to maintain a fixed number of significant digits (which is dictated by the 'word length' of the machine). For example if a computing machine retains exactly 6 significant decimal digits then a multiplication might be performed as follows:

.654321 × 1.00001 = .654328.

Since the true multiplication should be

.654321 + 1.00001 = .65432754321

a rounding error of about .0000005, or five in the seventh figure, is committed. In practice computers use *binary* arithmetic, but the effect is similar.

The sum total of all rounding errors committed up to a particular point in the calculation is termed the *accumulated rounding error*. If this accumulated rounding error grows relative to the computed values of the solution as we increase the number of values required, then the method of solution is termed *unstable*. Rounding errors are an important consideration in the direct methods of Chapter 5, but they are generally of little consequence in the iterative methods of Chapter 6 where the calculation is effectively restarted at each iteration. Stability is, however, seldom discussed in connection with the direct methods of Chapter 5, and this topic is most relevant in connection with other problems such as differential equations and recurrence relations.

*Significance error* is a serious form of rounding error, resulting from the subtraction of two nearly equal numbers. For example, using 6 significant decimal arithmetic, the calculation

6.54321 − 6.54320 = 0.00001

leads to a solution correct to at most *one* significant figure. Here the true calculation might be

6.543214 − 6.543196 = 0.000018

and in that case the percentage error committed would be about forty-four per cent.


## 2.3 Well-posed problems

It must also be pointed out that we should not embark on a problem in the first place unless we have ascertained that it is a well-posed problem, namely a problem having a unique and meaningful solution. For example if a rectangular system of $m$ ($>n$) linear algebraic equations in $n$ unknowns does not reduce to a set of $n$ equations, then no solutions exist and we have an ill-posed problem. On the other hand, this may be *approximately* replaced by a well-posed problem (see Section 3.11 below). Note that even a system $\mathbf{Ax} = \mathbf{b}$ of $n$ equations in $n$ unknowns may be ill-posed if the matrix $\mathbf{A}$ is singular (when there are no solutions unless $\mathbf{b} = \mathbf{0}$) or if $\mathbf{b} = \mathbf{0}$ (when there are no solutions unless $\mathbf{A}$ is singular). If both $\mathbf{b} = \mathbf{0}$ and $\mathbf{A}$ is singular, then the system is still ill-posed since there are then many solutions.

## 2.4 Floating point calculations

Calculations are normally performed with a fixed number of significant figures rather than a fixed number of decimal (or binary) places, and such arithmetic is termed *floating point arithmetic*. Adopting decimal rather than binary arithmetic, for argument's sake, this means that a number such as 1234.56 is represented as .123456 E4 (i.e. $.123456 \times 10^4$). Here the digits are shifted to left or right to bring the first non-zero digit into the first decimal place, resulting in a newly scaled number (.123456 in this case) called the *mantissa*, which must then be multiplied by a certain power of 10 (4 in this case) called the *exponent*. Arithmetic operations are still reasonably easy to organise and, for example, a multiplication might be performed as follows:

$$6543.21 \times .014 = .654321 \text{ E4} \times .140000 \text{ E}-1$$
$$= .0916049 \text{ E3}$$
$$= .916049 \text{ E2}$$

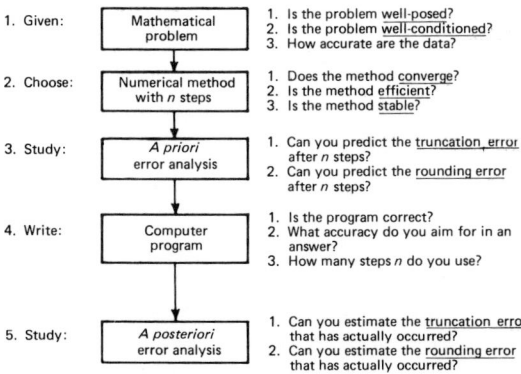

| 1. Given: | Mathematical problem | 1. Is the problem well-posed?<br>2. Is the problem well-conditioned?<br>3. How accurate are the data? |
| 2. Choose: | Numerical method with $n$ steps | 1. Does the method converge?<br>2. Is the method efficient?<br>3. Is the method stable? |
| 3. Study: | A priori error analysis | 1. Can you predict the truncation error after $n$ steps?<br>2. Can you predict the rounding error after $n$ steps? |
| 4. Write: | Computer program | 1. Is the program correct?<br>2. What accuracy do you aim for in an answer?<br>3. How many steps $n$ do you use? |
| 5. Study: | A posteriori error analysis | 1. Can you estimate the truncation error that has actually occurred?<br>2. Can you estimate the rounding error that has actually occurred? |

*Figure* 2.1. Steps in an ideal numerical method.

## 2.5 Summary

Figure 2.1 provides a list of the various steps that constitute an ideal numerical method. Note that error analysis has been included, both before the calculation (*a priori*) and after it (*a posteriori*).

## 2.6 References

1. Mason, J.C., *BASIC Numerical Mathematics*, Butterworths (1983).

# Chapter 3

# Introduction to matrices

## ESSENTIAL THEORY

What, you might well ask, are matrices, and why do we need them? A matrix is fundamentally a piece of notation. As such it is an attractive way of dressing up linear algebra so as to make it tidier and more manageable. Indeed a single symbol can be used to replace a whole table of information, and hence tremendous savings of space and effort are possible in the manipulation of sets of algebraic expressions. Sceptics, who question the need for matrices, might however refer to Lewis Carroll for support: 'I could have done it in a much more complicated way,' said the Red Queen, immensely proud.

The fundamental subject in which matrices are used, namely linear algebra, encompasses systems of linear equations, linear transformations of variables, linear spaces of functions and data, and so on. We shall mainly be concerned with the solution of linear equations (in Chapters 5 and 6), since this is a problem of great importance, and we shall also study 'eigenvalue problems' (in Chapter 6) and linear approximation to functions and data (in Chapter 7). The latter topic can be studied rather rigorously in the context of linear spaces, but we shall keep the discussion on a lighter level.

## 3.1 Definitions

A matrix $\mathbf{A}$ is a set of constants or variables arranged in a table of rows and columns. Thus a matrix of $m$ rows and $n$ columns, which is also termed '$m$ by $n$' or '$m \times n$', has the form

$$\mathbf{A} = \begin{bmatrix} a_{11} & a_{12} & \ldots & a_{1n} \\ a_{21} & a_{22} & \ldots & a_{2n} \\ \vdots & & & \vdots \\ a_{m1} & a_{m2} & \ldots & a_{mn} \end{bmatrix}$$

where $a_{11}, a_{12}, \ldots, a_{mn}$ are called the 'elements of $\mathbf{A}$' or 'entries in $\mathbf{A}$'. The entry $a_{ij}$ is sometimes also written $(\mathbf{A})_{ij}$. The pair of subscripts $i$ and $j$ attached to the element $a_{ij}$ denote, respectively, its row and column numbers ($i = 1, \ldots, m; j = 1, \ldots, n$). A matrix with the same number $n$ of rows as columns is called 'square' and 'of order $n$'. A matrix which is not necessarily square is called 'rectangular'.

The elements $a_{11}, a_{22}, \ldots, a_{ii}, \ldots$ are called the 'diagonal' elements, and all other elements are called 'off-diagonal' elements.

A special example of a matrix is a 'vector' (or more specifically a 'column vector'), which is a matrix of just one column such as

$$\mathbf{x} = \begin{bmatrix} x_1 \\ x_2 \\ \vdots \\ x_n \end{bmatrix} \quad \text{or} \quad \mathbf{b} = \begin{bmatrix} b_1 \\ b_2 \\ \vdots \\ b_n \end{bmatrix}$$

In this case only one subscript is required, and a typical element of $\mathbf{x}$ is its '$i$th component' $x_i$ appearing in row $i$. Note that we generally use small letters for vectors such as $\mathbf{x}$ and capital letters for matrices such as $\mathbf{A}$. We may similarly define 'row vectors' to be matrices with just one row.

It is sometimes useful to consider individual rows and columns of a rectangular matrix $\mathbf{A}$, and so we define $R_i(\mathbf{A})$ and $C_j(\mathbf{A})$ to denote row $i$ and column $j$ of $\mathbf{A}$ respectively. Thus $R_i(\mathbf{A})$ and $C_j(\mathbf{A})$ are a 'row vector' and 'column vector' given by

$$R_i(\mathbf{A}) = (a_{i1}\, a_{i2} \ldots a_{in}), \quad C_j(\mathbf{A}) = \begin{bmatrix} a_{1j} \\ a_{2j} \\ \vdots \\ a_{mj} \end{bmatrix}$$

## 3.2 Sums, and zero matrices

If two matrices $\mathbf{A}$ and $\mathbf{B}$ have the same number of rows and columns say $m$ rows and $n$ columns, then the sum $\mathbf{A} + \mathbf{B}$ is simply defined to be the $m \times n$ matrix $\mathbf{C}$ with elements

$$c_{ij} = a_{ij} + b_{ij} \qquad (i = 1, \ldots, m; j = 1, \ldots, n)$$

Similarly the difference $\mathbf{A} - \mathbf{B}$ is the $m \times n$ matrix $\mathbf{E}$ with elements

$$e_{ij} = a_{ij} - b_{ij} \qquad (i = 1, \ldots, m; j = 1, \ldots, n) \tag{3.1}$$

It is now appropriate to define the $m \times n$ *zero matrix* $\mathbf{O}_{mn}$ (or $\mathbf{O}$ for short) to be the $m \times n$ matrix whose elements are all zero. For it then follows from (3.1) that

$$\mathbf{A} - \mathbf{A} = \mathbf{O}_{mn} = \mathbf{O}$$

Note also that it is consistent to define a *scalar multiple* of a matrix

$$\mathbf{B} = \lambda \mathbf{A} \qquad (\lambda \text{ constant})$$

to be the matrix whose elements are *all* $\lambda$ times the corresponding elements of $\mathbf{A}$,

i.e. $\qquad\qquad b_{ij} = \lambda a_{ij} \qquad (i = 1, \ldots, m; j = 1, \ldots, n)$

## 3.3 Products, and identity matrices

Suppose that $\mathbf{A}$ and $\mathbf{B}$ are, respectively, given $m \times n$ and $n \times p$ matrices (so that the number $n$ of columns in $\mathbf{A}$ is equal to the number of rows in $\mathbf{B}$). Then the product $\mathbf{AB}$ of $\mathbf{A}$ and $\mathbf{B}$ is defined to be the $m \times p$ matrix $\mathbf{C}$, whose elements $c_{ij}$ are given by

$$c_{ij} = \sum_{k=1}^{n} a_{ik} b_{kj} \qquad (i = 1, \ldots, m; j = 1, \ldots, p) \qquad (3.2)$$

Note that if the number of columns in $\mathbf{A}$ is *not* equal to the number of rows in $\mathbf{B}$, then $\mathbf{AB}$ is not defined and we say that '$\mathbf{A}$ and $\mathbf{B}$ are not conformable for multiplication'.

In the particular case $p = 1$, $\mathbf{B}$ becomes an $n \times 1$ vector ($\mathbf{x}$ say). To be consistent with (3.2), we therefore define the product $\mathbf{Ax}$ of an $m \times n$ matrix $\mathbf{A}$ and an $n \times 1$ vector $\mathbf{x}$ to be an $m \times 1$ vector $\mathbf{y}$, whose elements $y_i$ are given by

$$y_i = \sum_{k=1}^{n} a_{ik} x_k \qquad (i = 1, \ldots, m) \qquad (3.3)$$

Note that in general the matrices $\mathbf{A}$, $\mathbf{B}$, and $\mathbf{C}$ (or $\mathbf{A}$, $\mathbf{x}$ and $\mathbf{y}$) are of different rectangular shapes. For example,

if $\mathbf{A} = \begin{bmatrix} 3 & 2 \\ 5 & -1 \\ 4 & 9 \end{bmatrix}$ and $\mathbf{B} = \begin{bmatrix} 1 & 5 & 7 & 2 \\ 4 & 8 & 1 & 6 \end{bmatrix}$,

then $\mathbf{AB} = \mathbf{C} = \begin{bmatrix} 11 & 31 & 23 & 18 \\ 1 & 17 & 34 & 4 \\ 40 & 92 & 37 & 62 \end{bmatrix}$

and if $\mathbf{A} = \begin{bmatrix} 3 & 2 \\ 5 & -1 \\ 4 & 9 \end{bmatrix}$ and $\mathbf{x} = \begin{bmatrix} 1 \\ 4 \end{bmatrix}$, then $\mathbf{Ax} = \mathbf{y} = \begin{bmatrix} 11 \\ 1 \\ 40 \end{bmatrix}$

Observe that $\mathbf{BA}$ and $\mathbf{xA}$ cannot be defined in these examples.

It is now natural to define the *identity matrix* $\mathbf{I}_n$ of order $n$ (or $\mathbf{I}$ for short) to be the square $n \times n$ matrix whose diagonal elements are ones and whose off-diagonal elements are zeros. Thus

$$I = I_n = \begin{bmatrix} 1 & 0 & 0 & \ldots & 0 & 0 & 0 \\ 0 & 1 & 0 & \ldots & 0 & 0 & 0 \\ 0 & 0 & 1 & \ldots & 0 & 0 & 0 \\ \vdots & \vdots & \vdots & & \vdots & \vdots & \vdots \\ 0 & 0 & 0 & \ldots & 0 & 1 & 0 \\ 0 & 0 & 0 & \ldots & 0 & 0 & 1 \end{bmatrix} \Biggr\} \ n \text{ rows} \qquad (3.4)$$

It is easy to verify from (3.4) that $I_n$ has the properties expected of an identity. Specifically if $A$ is $m \times n$ and $B$ is $n \times p$, then

$$AI_n = A \qquad \text{and} \qquad I_n B = B$$

These properties simplify for a square $n \times n$ matrix $A$ to

$$AI_n = I_n A = A.$$

Note from the above definitions that $AB$ and $BA$ are not necessarily both defined. If they are, then $A$ and $B$ must both be square matrices of the same order, say $n$. However, even then it does *not* follow that $AB$ and $BA$ are equal. For example,

if $A = \begin{bmatrix} 1 & 2 \\ 3 & 4 \end{bmatrix}$ and $B = \begin{bmatrix} 1 & -1 \\ 1 & 0 \end{bmatrix}$, then $AB = \begin{bmatrix} 3 & -1 \\ 7 & -3 \end{bmatrix}$ and

$$BA = \begin{bmatrix} -2 & -2 \\ 1 & 2 \end{bmatrix}$$

We finally remark that the definition (3.2) of $AB$ may be replaced by a 'simpler' definition, based on the scalar (i.e. 1 by 1) product of a row vector and a column vector

$$(x_1 \, x_2 \ldots x_n) \begin{bmatrix} y_1 \\ y_2 \\ \vdots \\ y_n \end{bmatrix} = x_1 y_1 + x_2 y_2 + \ldots + x_n y_n = \sum_{k=1}^{n} x_k y_k \qquad (3.5)$$

Note that this is exactly the definition for the scalar product of a pair of algebraic or geometric vectors $x = (x_1, x_2, \ldots, x_n)$ and $y = (y_1, y_2, \ldots, y_n)$. Then (3.2) is equivalent to

$$c_{ij} = (AB)_{ij} = R_i(A) . C_j(B) \qquad (3.6)$$

i.e. the element in row $i$ and column $j$ of $AB$ is obtained by multiplying row $i$ of $A$ by column $j$ of $B$.

### 3.4 Transposes, symmetric and positive definite matrices

The 'transpose matrix' $A^T$ (sometimes written $A'$) of a matrix $A$ is that matrix whose rows are the columns of $A$ (and whose columns are the rows of $\mathbf{A}$). Algebraically, the elements of the matrix $\mathbf{A}^T$ are related to those of $\mathbf{A}$ by

$$(\mathbf{A})_{ij} = (\mathbf{A}^T)_{ji} \tag{3.7}$$

Thus if $\mathbf{A}$ is $m \times n$, then $\mathbf{A}^T$ is $n \times m$ and for example

$$\begin{pmatrix} 1 & 4 \\ 2 & 5 \\ 3 & 6 \end{pmatrix}^T = \begin{pmatrix} 1 & 2 & 3 \\ 4 & 5 & 6 \end{pmatrix}$$

The transpose of a product of two matrices may easily be seen to be the reversed product of the transposed matrices,

i.e. $$(\mathbf{AB})^T = \mathbf{B}^T\mathbf{A}^T \tag{3.8}$$

For by (3.5) and (3.6),

$$(\mathbf{AB})_{ij} = R_i(\mathbf{A})C_j(\mathbf{B}) = R_j(\mathbf{B}^T)C_i(\mathbf{A}^T) = (\mathbf{B}^T\mathbf{A}^T)_{ji}$$

and hence the result follows by (3.7).

One simple use of a transpose is for writing a column vector on one line. For example

$$\begin{pmatrix} x_1 \\ x_2 \\ x_3 \end{pmatrix}$$

may be written as $(x_1 \ x_2 \ x_3)^T$.

An important class of matrices are *symmetric* matrices $\mathbf{A}$ with the property that

$$(\mathbf{A})_{ij} = (\mathbf{A})_{ji} \tag{3.9}$$

Such matrices have equal elements above and below the diagonal elements in corresponding positions. In terms of transpose matrices, this means that

$$\mathbf{A} = \mathbf{A}^T \tag{3.10}$$

One interesting application of transpose matrices is in the efficient representation of *quadratic forms*. For example, if $\mathbf{A}$ is a given $3 \times 3$ matrix with elements $a_{ij}$, and $\mathbf{x}$ is a $3 \times 1$ vector with elements $x_i$, then

$$\mathbf{x}^T\mathbf{A}\mathbf{x} = (x_1 \ x_2 \ x_3)\begin{pmatrix} a_{11} & a_{12} & a_{13} \\ a_{21} & a_{22} & a_{23} \\ a_{31} & a_{32} & a_{33} \end{pmatrix}\begin{pmatrix} x_1 \\ x_2 \\ x_3 \end{pmatrix}$$

$$= a_{11}x_1{}^2 + a_{22}x_2{}^2 + a_{33}x_3{}^2 + (a_{23} + a_{32})x_2x_3$$
$$+ (a_{31} + a_{13})x_3x_1 + (a_{12} + a_{21})x_1x_2$$

$$= \sum_{i=1}^{3} \sum_{j=1}^{3} a_{ij}x_ix_j$$

Clearly this is a quadratic in $x_1$, $x_2$, $x_3$. More generally, if **A** is a given $n \times n$ matrix and **x** is a given $n \times 1$ vector, we have the identity

$$\mathbf{x}^{\mathrm{T}}\mathbf{A}\mathbf{x} = \sum_{i=1}^{n} \sum_{j=1}^{n} a_{ij}x_ix_j \qquad (3.11)$$

and the right-hand side is the most general form for a quadratic in $x_1, x_2, \ldots, x_n$.

Conversely, we may express a given quadratic in matrix form. However, we then have many choices of **A** available to us, and indeed it is always possible to choose **A** to be symmetric. For example

$$x_1{}^2 + x_2{}^2 - x_3{}^3 + 6x_2x_3 - 4x_3x_1 + 2x_1x_2$$
$$= x_1{}^2 + x_2{}^2 - x_3{}^3 + (3x_2x_3 + 3x_3x_2) + (-2x_3x_1 - 2x_1x_3)$$
$$+ (x_1x_2 + x_2x_1)$$
$$= \mathbf{x}^{\mathrm{T}}\mathbf{A}\mathbf{x}, \text{ where } \mathbf{x} = \begin{bmatrix} x_1 \\ x_2 \\ x_3 \end{bmatrix} \text{ and } \mathbf{A} = \begin{bmatrix} 1 & 1 & -2 \\ 1 & 1 & 3 \\ -2 & 3 & -1 \end{bmatrix}$$

(The reason that we have more than one choice of **A** is that we may choose any pair of values for $a_{23}$ and $a_{32}$ such that $a_{23} + a_{32} = 6$, and we have similar latitude with $a_{31}$ and $a_{13}$, and with $a_{12}$ and $a_{21}$. If we choose $a_{23} = a_{32}$, $a_{31} = a_{13}$, $a_{12} = a_{21}$, then **A** becomes symmetric.)

The compact matrix formula (3.11) for a quadratic enables us to introduce the idea of a matrix **A** which is *positive definite*. Such a matrix is one for which $\mathbf{x}^{\mathrm{T}}\mathbf{A}\mathbf{x} \geqslant 0$ for *all* **x**, with equality if and only if $\mathbf{x} = \mathbf{O}$. A simple example of a positive definite $3 \times 3$ matrix is the identity matrix $\mathbf{I}_3$, since, for a $3 \times 1$ vector **x**,

$$\mathbf{x}^{\mathrm{T}}\mathbf{I}\mathbf{x} = x_1{}^2 + x_2{}^2 + x_3{}^2$$

and clearly this quadratic is zero if and only if $x_1 = x_2 = x_3 = 0$ (and hence $\mathbf{x} = \mathbf{O}$).

### 3.5 Linear transformations

A matrix **A** can also be regarded as representing a *linear transformation* (or linear change) of variables

$$\mathbf{y} = \mathbf{Ax} \tag{3.12}$$

from the vector variable $\mathbf{x}$ to the vector variable $\mathbf{y}$. For example, in two dimensions, if our two scalar variables are $x_1$ and $x_2$ (in place of the conventional $x$ and $y$), then we may effectively rotate the axes through $45°$ and consider new scalar variables $y_1$ and $y_2$, by making the linear transformation

$$y_1 = x_1 + x_2 \qquad y_2 = x_1 - x_2$$

(The new axes $y_2 = 0$ and $y_1 = 0$ are the old lines $x_1 = \pm x_2$, which are at $45°$ to the old axes $x_2 = 0$ and $x_1 = 0$.)

Then
$$\mathbf{y} = \begin{bmatrix} y_1 \\ y_2 \end{bmatrix} = \begin{bmatrix} 1 & 1 \\ 1 & -1 \end{bmatrix} \begin{bmatrix} x_1 \\ x_2 \end{bmatrix} = \mathbf{Ax}$$

where $\mathbf{A} = \begin{bmatrix} 1 & 1 \\ 1 & -1 \end{bmatrix}$ represents the linear transformation.

More generally, the $n \times n$ matrix $\mathbf{A}$ with elements $a_{ij}$ represents the transformation of variables from $\mathbf{x} = (x_1 \, x_2 \ldots x_n)^{\mathrm{T}}$ to $\mathbf{y} = (y_1 \, y_2 \ldots y_n)^{\mathrm{T}}$ given by (3.12). Clearly this means that

$$y_i = (\mathbf{Ax})_i = \sum_{j=1}^{n} a_{ij} x_j = a_{i1} x_1 + a_{i2} x_2 + \ldots + a_{in} x_n \tag{3.13}$$
$$(i = 1, 2, \ldots, n)$$

Hence (3.12) is equivalent to the set of equations

$$\left. \begin{aligned} y_1 &= a_{11} x_1 + a_{12} x_2 + \ldots + a_{1n} x_n \\ y_2 &= a_{21} x_1 + a_{22} x_2 + \ldots + a_{2n} x_n \\ &\vdots \\ y_n &= a_{n1} x_1 + a_{n2} x_2 + \ldots + a_{nn} x_n \end{aligned} \right\} \tag{3.14}$$

## 3.6 Determinants, and cofactors

The determinant, written $|\mathbf{A}|$ or det $(\mathbf{A})$, of a matrix $\mathbf{A}$ is a scalar quantity calculated by a repetitive process. (There is not enough space here to go into a long discussion of determinants, and the reader is referred to Reference 3 for more details.)

The determinant of an $n \times n$ matrix $\mathbf{A}$ is defined in terms of the determinants of a number of $(n - 1) \times (n - 1)$ matrices. Each of these may be defined in terms of determinants of $(n - 2) \times (n - 2)$ matrices, and so on. Finally the determinant of a $1 \times 1$ matrix is required, and this is just defined to be the (one) element in the matrix. We shall 'expand' the determinant along the 1st row (although in fact any row or column may be used).

The *cofactor* $A_{ij}$ of an element $a_{ij}$ in the $n \times n$ matrix **A** is defined to be the determinant of the $(n-1) \times (n-1)$ matrix formed by deleting row $i$ and column $j$ of **A**. Then the *determinant* of **A** is defined by:

$$|\mathbf{A}| = a_{11}A_{11} - a_{12}A_{12} + a_{13}A_{13} - \ldots + (-1)^{n-1}a_{1n}A_{1n}$$

For example, suppose that $\mathbf{A} = \begin{pmatrix} 1 & 2 & 3 \\ 4 & 5 & 4 \\ 3 & 2 & 1 \end{pmatrix}$

Then $|\mathbf{A}| = 1. \begin{vmatrix} 5 & 4 \\ 2 & 1 \end{vmatrix} - 2. \begin{vmatrix} 4 & 4 \\ 3 & 1 \end{vmatrix} + 3. \begin{vmatrix} 4 & 5 \\ 3 & 2 \end{vmatrix}$

where $\begin{vmatrix} 5 & 4 \\ 2 & 1 \end{vmatrix} = 5.|1| - 4.|2| = -3, \quad \begin{vmatrix} 4 & 4 \\ 3 & 1 \end{vmatrix} = 4.|1| - 4.|3|$
$$= -8$$

and $\begin{vmatrix} 4 & 5 \\ 3 & 2 \end{vmatrix} = 4.|2| - 5.|3| = -7.$

Hence $\qquad |\mathbf{A}| = 1.(-3) - 2.(-8) + 3.(-7) = -8$

### 3.7 Inverses

Having defined a (square) identity matrix $\mathbf{I}_n$ of order $n$, it is easy to introduce in principle the idea, for any given square $n \times n$ matrix **A**, of an $n \times n$ matrix **B** with the property that

$$\mathbf{AB} = \mathbf{BA} = \mathbf{I}$$

In some sense **B** must be the 'reciprocal' of **A**, and in fact we call it the *inverse matrix* $\mathbf{A}^{-1}$ of order $n$ such that

$$\mathbf{AA}^{-1} = \mathbf{A}^{-1}\mathbf{A} = \mathbf{I}$$

For example, it is easily verified (for $n = 2$) that

if $\qquad \mathbf{A} = \begin{pmatrix} 1 & 1 \\ 1 & -1 \end{pmatrix}$ then $\mathbf{A}^{-1} = \begin{pmatrix} .5 & .5 \\ .5 & -.5 \end{pmatrix}$ $\qquad$ (3.15)

In fact there is a 'simple' formula for $\mathbf{A}^{-1}$, for any given **A**, based on the cofactors $A_{ij}$ (see Reference 3 for a derivation). Let us first define the *adjugate matrix* adj **A** of **A** to be the transposed matrix formed from the cofactors $A_{ij}$ with (alternating) signs $(-1)^{i+j}$ attached to these entries:

adj $\mathbf{A} =$ (3.16)

$$= \begin{bmatrix} +A_{11} & -A_{12} & +A_{13} & \cdots & (-1)^{n+1}A_{1n} \\ -A_{21} & +A_{22} & -A_{23} & \cdots & (-1)^{n+2}A_{2n} \\ +A_{31} & -A_{32} & +A_{33} & \cdots & (-1)^{n+3}A_{3n} \\ \vdots & & & & \vdots \\ (-1)^{n+1}A_{n1} & (-1)^{n+2}A_{n2} & (-1)^{n+3}A_{n3} & \cdots & (-1)^{2n}A_{nn} \end{bmatrix}^{T}$$

Then $$\mathbf{A}^{-1} = (\text{adj } \mathbf{A})/|\mathbf{A}|$$ (3.17)

If $|\mathbf{A}| = 0$, then $\mathbf{A}^{-1}$ is *not defined*, and we say that $\mathbf{A}$ is *singular*. For example, in (3.15)

$$\text{adj } \mathbf{A} = \begin{bmatrix} +(-1) & -(1) \\ -(1) & +(1) \end{bmatrix}^{T} = \begin{bmatrix} -1 & -1 \\ -1 & 1 \end{bmatrix} \text{ and } |\mathbf{A}| = -2$$

However, if $\mathbf{A} = \begin{bmatrix} 1 & 1 \\ 2 & 2 \end{bmatrix}$, then $|\mathbf{A}| = 0$ and $\mathbf{A}$ is singular.

[In the above discussion the 'ratio' $\mathbf{A}/\lambda$ stands for the matrix $\lambda^{-1}\mathbf{A}$ whose entries are those of $\mathbf{A}$ each divided by $\lambda$.]

### 3.8 Linear equations

By far the most important application of matrices is in the solution of a set of simultaneous linear equations. A system of $n$ equations in $n$ unknowns $x_1, x_2, \ldots, x_n$ takes the form

$$\left. \begin{array}{l} a_{11}x_1 + a_{12}x_2 + \ldots + a_{1n}x_n = b_1 \\ a_{21}x_1 + a_{22}x_2 + \ldots + a_{2n}x_n = b_2 \\ \vdots \qquad\qquad\qquad\qquad \vdots \\ a_{n1}x_1 + a_{n2}x_2 + \ldots + a_{nn}x_n = b_n \end{array} \right\}$$ (3.18)

where $a_{11}, \ldots, a_{nn}, b_1, \ldots, b_n$ are given constants.

If $\mathbf{A}$ is an $n \times n$ (known) matrix with elements $a_{ij}$, $\mathbf{b}$ is an $n \times 1$ (known) vector with elements $b_i$, and $\mathbf{x}$ is an $n \times 1$ unknown vector with elements $x_i$, then it follows immediately from (3.3) that (3.18) may be rewritten as a single matrix equation

$$\mathbf{Ax} = \mathbf{b}$$ (3.19)

From Section 3.7 the solution of (3.19) may immediately be written down in the unique form

$$\mathbf{x} = \mathbf{A}^{-1}\mathbf{b}$$ (3.20)

provided $|\mathbf{A}| \neq 0$.

In case $|\mathbf{A}| = 0$, there are two possibilities. If $\mathbf{b} = \mathbf{O}$, then all right hand sides are zero. We then have a *homogeneous* system of equations, and there are typically *many solutions* for $\mathbf{x}$. A standard idea then is to fix one component, say $x_1$, of $\mathbf{x}$ and try to find unique values of $x_2, \ldots, x_n$ by solving only $n - 1$ of the equations. However, if $\mathbf{b} \neq \mathbf{O}$, then there may be either *many solutions* or *no solutions* for $\mathbf{x}$.

For example, $\left. \begin{array}{l} x_1 + x_2 = 2 \\ 2x_1 + 2x_2 = 4 \end{array} \right\}$ have many solutions

and $\left. \begin{array}{l} x_2 + x_2 = 2 \\ 2x_1 + 2x_2 = 5 \end{array} \right\}$ have no solutions.

(The complete solution, covering all possible cases, of a pair of simultaneous equations with integer coefficients is discussed in Reference 1.)

## 3.9  Eigenvalue problems

The second most important application is probably that of determining, for a given matrix $\mathbf{A}$, a vector $\mathbf{x}$ (called an *eigenvector*) and corresponding scalar $\lambda$ (called an *eigenvalue*) with the property that

$$\mathbf{A}\mathbf{x} = \lambda \mathbf{x} \tag{3.21}$$

For an $n \times n$ matrix $\mathbf{A}$, there are generally $n$ possible eigenvalues $\lambda$ and $n$ corresponding eigenvectors $\mathbf{x}$. We shall not discuss the solution of eigenvalue problems here, but both mathematical and numerical mathematical methods will be discussed in Chapter 5.

## 3.10  Collocation

In this and the following section we refer briefly to two important applications of matrix methods in approximation and data fitting, which will be discussed in detail in Chapter 7.

If a given function $f(x)$ and a polynomial

$$f^*(x) = c_1 + c_2 x + c_3 x^2 + \ldots + c_n x^{n-1}$$

are set equal at $n$ distinct $x$ values $x_1, x_2, \ldots, x_n$, then we obtain the set of $n$ simultaneous linear equations

$$c_1 + c_2 x_i + c_3 (x_i)^2 + \ldots + c_n (x_i)^{n-1} = y_i = f(x_i) \qquad (i = 1, \ldots, n)$$

This may be written in the matrix form:

$$\mathbf{A}\mathbf{c} = \mathbf{y} \tag{3.22}$$

for the determination of $\mathbf{c} = (c_1 c_2 \ldots c_n)^{\mathrm{T}}$ from $\mathbf{y} = (y_1 y_2 \ldots y_n)^{\mathrm{T}}$,

where 
$$\mathbf{A} = \begin{bmatrix} 1 & x_1 & (x_1)^2 & \cdots & (x_1)^{n-1} \\ 1 & x_2 & (x_2)^2 & \cdots & (x_2)^{n-1} \\ \vdots & & \vdots & & \vdots \\ 1 & x_n & (x_n)^2 & \cdots & (x_n)^{n-1} \end{bmatrix} \qquad (3.23)$$

It is not difficult to see, by subtracting rows and extracting factors $x_1 - x_2$, etc, that

$$|\mathbf{A}| = \lambda . \prod_{i \neq j} (x_i - x_j) \qquad (3.24)$$

where the product is taken over all pairs $i$ and $j$ from 1 to $n$, excepting $i = j$, and where the constant $\lambda$ is $\pm 1$ (depending on the ordering of the pairs $x_i$, $x_j$).

For example, for $n = 3$, $\qquad |\mathbf{A}| = (x_2 - x_3)(x_3 - x_1)(x_1 - x_2)$

It follows immediately from (3.24) that, since $x_1, \ldots, x_n$ are all distinct, $|\mathbf{A}| \neq 0$. Hence the *collocation equations* (3.22) have a *unique solution* (provided that $f(x_1), f(x_2), \ldots, f(x_n)$ are not all zero).

### 3.11 Least squares

If in Section 3.10, $f(x)$ is equated to $f^*(x)$ at $m > n$ points $x_1$, $x_2$, $\ldots$, $x_m$, then we obtain a 'rectangular' system of linear equations

$$\mathbf{Bc} = \mathbf{y} \qquad (3.25)$$

where $\mathbf{B}$ is a rectangular $m \times n$ matrix, $\mathbf{y}$ is an $m \times 1$ matrix, and $\mathbf{c}$ is the required $n \times 1$ matrix of coefficients. (Here $\mathbf{B}$ is the same matrix as $\mathbf{A}$ in Section 3.10, except that it now has $m$ rows.)

The system (3.25) is *over-determined*, since it involves more equations than unknowns, and generally no solution exists. However, all is not lost, and in Chapter 7 we shall see that, by using the *method of least squares*, we may solve (3.25) approximately by multiplying both sides by $\mathbf{B}^T$. This leads to the *well-determined* system of equations

$$\mathbf{Ac} = \mathbf{b} \qquad (3.26)$$

where $\qquad \mathbf{A} = \mathbf{B}^T\mathbf{B} \quad$ and $\quad \mathbf{b} = \mathbf{B}^T\mathbf{y}$

Note that $\mathbf{A}$ is now a square $n \times n$ matrix (called the *normal matrix* of $\mathbf{B}$), and $\mathbf{b}$ is an $n \times 1$ vector. Moreover it can be shown that $|\mathbf{A}| \neq 0$, and hence that (3.26) has a unique solution.

The normal matrix $\mathbf{A}$ has two desirable properties. Firstly, it is symmetric, since

$$\mathbf{A}^T = (\mathbf{B}^T\mathbf{B})^T = \mathbf{B}^T\mathbf{B}^{TT} = \mathbf{B}^T\mathbf{B} = \mathbf{A}$$

Secondly, it is positive definite. For, suppose that $\mathbf{x}$ is any $n \times 1$

vector, and that $\mathbf{z}$ is the $m \times 1$ vector $\mathbf{Bx}$ with components $z_1$, $z_2$, ..., $z_m$. Then

$$\mathbf{x}^T\mathbf{A}\mathbf{x} = \mathbf{x}^T\mathbf{B}^T\mathbf{B}\mathbf{x} = (\mathbf{B}\mathbf{x})^T(\mathbf{B}\mathbf{x}) = \mathbf{z}^T\mathbf{z} = z_1{}^2 + z_2{}^2 + \ldots + z_m{}^2 \geq 0$$

with equality if and only $\mathbf{z} = 0$ (which only occurs if $\mathbf{x} = 0$).

## 3.12 Further reading

A good introductory treatment of matrices (in rather more depth than ours) may be found in Reference 2, and a more detailed and elementary discussion may be found in a 'traditional mathematical' text such as Reference 3.

## 3.13 References

1. Mason, J.C. *BASIC Numerical Mathematics*, Butterworths (1983).
2. Fox, L. *Introduction to Numerical Linear Algebra*, Oxford (1964).
3. Grossman, S.I. *Elementary Linear Algebra*, Wadsworth (International Student Edition) (1980).

## PROBLEMS

Set $\mathbf{A} = \begin{bmatrix} 3 & -2 & 1 \\ -2 & 6 & 0 \\ 1 & 0 & 4 \end{bmatrix}$, $\mathbf{B} = \begin{bmatrix} 1 & 0 & 1 \\ 0 & 1 & 0 \\ 1 & 2 & 1 \end{bmatrix}$, $\mathbf{C} = \begin{bmatrix} 1 & 2 \\ 3 & 4 \\ 1 & 3 \end{bmatrix}$,

$$\mathbf{D} = \begin{bmatrix} 1 & 0 \\ 1 & 1 \end{bmatrix}$$

**(3.1)** Calculate $\mathbf{A} + \mathbf{B}$, $\mathbf{A} - \mathbf{B}$, $2\mathbf{A} - 5\mathbf{B}$.
Are any of the sums and differences $\mathbf{B} \pm \mathbf{C}$, $\mathbf{C} \pm \mathbf{D}$ defined?
**(3.2)** Verify that $\mathbf{AB} \neq \mathbf{BA}$.
Calculate any of the products $\mathbf{BC}$, $\mathbf{CB}$, $\mathbf{CD}$, $\mathbf{DC}$ which are defined.
**(3.3)** Determine $\mathbf{A}^T$, $\mathbf{B}^T$, and $(\mathbf{AB})^T$, and verify that $(\mathbf{AB})^T = \mathbf{B}^T\mathbf{A}^T$.
**(3.4)** Are either $\mathbf{A}$ or $\mathbf{B}$ symmetric?
Calculate the matrix $\mathbf{B}^T\mathbf{B}$ and verify that it is symmetric.
**(3.5)** If $\mathbf{x} = (x_1 x_2 x_3)^T$, verify that
$$\mathbf{x}^T\mathbf{A}\mathbf{x} = x_1{}^2 + 2x_2{}^2 + 3x_3{}^2 + (x_1 - 2x_2)^2 + (x_1 + x_3)^2$$
and deduce that $\mathbf{A}$ is positive definite.
**(3.6)** Use formula (3.17) to determine the matrix $\mathbf{A}^{-1}$, and check your result by calculating $\mathbf{A}^{-1}\mathbf{A}$ and $\mathbf{A}\mathbf{A}^{-1}$.
**(3.7)** If $\mathbf{y}$ is obtained from $\mathbf{x}$ by the linear transformation $\mathbf{y} = \mathbf{Ax}$ (where $\mathbf{x}$ and $\mathbf{y}$ are $3 \times 1$ matrices), find a matrix $\mathbf{E}$ such that $\mathbf{x}$ is obtained from $\mathbf{y}$ by the linear transformation $\mathbf{x} = \mathbf{Ey}$. (See Problem 6.)

If $y = Ax$ and $x = Bz$, find the matrix $F$ for the linear transformation $y = Fz$. (See Problem 2.) Note: This is an alternative way of defining inverses and products of matrices.

(3.8) If $b = (1\ 1\ 1)^T$, use $A^{-1}$ from Problem 6 to solve for $x$ the system of linear equations $Ax = b$.
Calculate $Ax - b$ based on the solution $x$ so obtained, and hence verify that $x$ satisfies the given equations.

(3.9) Prove the formula (3.24) for the determinant of the 'collocation matrix' (3.23). Determine the sign of $\lambda (= \pm 1)$ for your ordering of $i$ and $j$ in the product (3.24) by equating terms in $1.x_2.x_3{}^2...(x_n)^{n-1}$ (i.e. the main diagonal term of $|A|$). Note: If $i < j$, for all $i$ and $j$, then $\lambda = (-1)^r$ where $r = 1 + 2 + ... + (n-1) = \frac{1}{2}n(n-1)$.

(3.10) Do Problem 11 of Chapter 5, on the least squares solution of a system of linear equations, based on the discussion of Section 3.11.

(3.11) You may assume that, for square matrices, $|A| = |A^T|$ and $|AB| = |A|.|B|$.
If a matrix $A$ satisfies $A^{-1} = A^T$ then it is said to be *orthogonal*.
Verify that the matrix $\begin{bmatrix} .6 & .8 \\ -.8 & .6 \end{bmatrix}$ is orthogonal.
Prove that if $A$ and $B$ are any orthogonal matrices of the same order, then (i) $AB$ is orthogonal (ii) $|A| = \pm 1$.

Chapter 4

# Elementary matrix calculations

## ESSENTIAL THEORY

In Table 1.1 we listed the matrix routines which are available as matrix statements in BASIC. All but one of these routines, the exception being the inversion $\mathbf{A} = \mathbf{B}^{-1}$, are very elementary calculations to program in BASIC. In this chapter we shall focus on these routines, not only because of their fundamental importance but also in order to give the reader simple practice in BASIC programming. The exceptional routine $\mathbf{A} = \mathbf{B}^{-1}$ is covered in Algorithm 5.4B of the next chapter.

## 4.1 Input and output

Throughout this book it will be assumed that a single terminal is used not only for entering programs but also for the input of data and the output of results. Moreover conversational programming based on INPUT statements (see Section 1.3.4) will generally be used, so that programs can run themselves and ask for any appropriate data that are required.

Before considering INPUT statements, let us note that READ statements permit the display of a matrix in its standard tableau form. For example, the following routine reads in numerical entries for an $m \times n$ matrix $\mathbf{A}$, where $m$ and $n$ do not exceed 10, the particular data in this case correspond to a $3 \times 3$ unit matrix.

```
 10 DIM A(10, 10)
 20 READ M, N
 30 FOR I = 1 TO M
 40 FOR J = 1 TO N
 50 READ A(I, J)
 60 NEXT J
 70 NEXT I
 80 DATA 3, 3
 90 DATA 1, 0, 0
100 DATA 0, 1, 0
110 DATA 0, 0, 1
```

(4.1)

On computers with dynamic dimensioning, we might make the efficient modification:

DELETE 10
25  DIM A(M, N)

Note that statements 30–70 are equivalent to the single matrix statement

50  MAT READ A

provided that A has the precise dimensions M and N. However, the latter matrix statement 50 would either have to be preceded by the statement 25 above or an equivalent statement. The two routines below carry out the same task as (4.1) using MAT statements, the left- and right-hand side routines being respectively appropriate to machines with and without dynamic dimensioning. Note that the second routine sets the precise dimensions M and N of A in statement 50.

| | |
|---|---|
| 20  READ M, N ⎫<br>25  DIM A(M, N) ⎪<br>50  MAT READ A ⎪<br>80  DATA 3, 3 ⎬ (4.2)<br>90  DATA 1, 0, 0 ⎪<br>100  DATA 0, 1, 0 ⎪<br>110  DATA 0, 0, 1 ⎭ | 10  DIM A(10, 10) ⎫<br>20  READ M, N ⎪<br>50  MAT READ A(M, N) ⎪<br>80  DATA 3, 3 ⎬ (4.3)<br>90  DATA 1, 0, 0 ⎪<br>100  DATA 0, 1, 0 ⎪<br>110  DATA 0, 0, 1 ⎭ |

In all the above routines the matrix **A** has been prescribed in DATA statements 90–110, one row per statement. Of course a single DATA statement could replace 90–110:

90  DATA 1, 0, 0, 0, 1, 0, 0, 0, 1.

In the case of INPUT statements, the data are specified at the keyboard. Corresponding routines to (4.1) and (4.3) might be the following ones. Routine (4.4) uses standard BASIC only, while (4.5) uses MAT statements.

| | |
|---|---|
| 10  DIM A(10, 10) ⎫<br>19  PRINT "M, N"; ⎪<br>20  INPUT M, N ⎪<br>29  PRINT "MATRIX A" ⎪<br>30  FOR I = 1 TO M ⎬ (4.4)<br>40  FOR J = 1 TO N ⎪<br>50  INPUT A(I, J) ⎪<br>60  NEXT J ⎪<br>70  NEXT I ⎭ | 10  DIM A(10, 10) ⎫<br>19  PRINT "M, N"; ⎪<br>20  INPUT M, N ⎬ (4.5)<br>29  PRINT "MATRIX A" ⎪<br>50  MAT INPUT A(M, N) ⎭ |

The statements 19 and 29 have been included in order to provide a message at the keyboard. For routine (4.4) the conversation is as follows:

| COMPUTER PRINTS: | USER REPLIES: |
|---|---|
| M, N? | 3, 3 |
| MATRIX A | |
| ? | 1 |
| ? | 0 |
| ? | 0 |
| ? | 0 |
| ? | 1 |
| ? | 0 |
| ? | 0 |
| ? | 0 |
| ? | 1 |

Here the INPUT statement 50 calls for the typing of just one entry of **A** per line, and so we cannot present a neat matrix format. On the other hand, for routine (4.5) the conversation is as follows:

| COMPUTER PRINTS: | USER REPLIES: |
|---|---|
| M, N? | 3, 3 |
| MATRIX A | |
| ? | 1, 0, 0, 0, 1, 0, 0, 0, 1 |

In this case the INPUT statement 50 calls for the typing on a single line of all elements of **A** listed by row. However, by inserting after each row the appropriate 'continuation character' ('&' on the VAX, for example), a standard matrix format for **A** may be achieved. For routine (4.5) the conversation with the computer then proceeds as follows:

| COMPUTER PRINTS: | USER REPLIES: |
|---|---|
| M, N? | 3, 3 |
| MATRIX A | |
| ? | 1, 0, 0, & |
| ? | 0, 1, 0, & |
| ? | 0, 0, 1. |

For output purposes, the following two routines might follow (4.4) and (4.5) respectively.

```
120 FOR I = 1 TO M
130 FOR J = 1 TO N
140 PRINT A(I, J)        (4.6)      140 MAT PRINT A      (4.7)
150 NEXT J
160 NEXT I
```

Both these routines will result in **A** being output one element per line. A more desirable output of **A** in rows can be achieved in (4.6) by making the following changes:

    140 PRINT A(I, J),
    155 PRINT

The comma in 140 gives a standard spacing. However, it is then essential that an appropriate *margin* should be fixed either on the keyboard or by the relevant BASIC statement (see the user manual for your machine), otherwise the output device may change lines in the middle of a number. The statement 155 starts a new line for each row of **A**.

## 4.2 Arithmetic statements by double loops

Apart from the matrix multiplication $\mathbf{A} = \mathbf{BC}$ (and $\mathbf{A} = \mathbf{B}^{-1}$ which we have already excluded), all remaining statements in Table 1.1 may be coded by using just two nested loops (i.e. two pairs of FOR and NEXT statements, one pair within the other), in much the same way as was done in Section 4.1. We give two examples below. Routine (4.8) calculates $\mathbf{A} = \mathbf{B}^{\mathrm{T}}$, where **B** is an $m \times n$ matrix, and routine (4.9) calculates $\mathbf{A} = \mathbf{I}$, where **I** is $n \times n$. It is assumed in (4.8) that **B** has already been input, and in both routines that **A** and **B** have been dimensioned.

$$
\left.\begin{array}{l}
\text{10  FOR I = 1 TO N} \\
\text{20  FOR J = 1 TO M} \\
\text{30  A(I, J) = B(J, I)} \\
\text{40  NEXT J} \\
\text{50  NEXT I}
\end{array}\right\}(4.8)
\qquad
\left.\begin{array}{l}
\text{10  FOR I = 1 TO N} \\
\text{20  FOR J = 1 TO N} \\
\text{30  A(I, J) = 0} \\
\text{40  NEXT J} \\
\text{50  A(I, J) = 1} \\
\text{60  NEXT I}
\end{array}\right\}(4.9)
$$

## 4.3 Matrix multiplication

Suppose that **B** is a given $m \times n$ matrix with elements $b_{ij}$, and that **C** is a given $n \times p$ matrix with elements $c_{ij}$, then the product $\mathbf{A} = \mathbf{BC}$ is an $m \times p$ matrix with elements (see Section 3.3)

$$
a_{ij} = \sum_{k=1}^{n} b_{ik} c_{kj} \quad (i = 1, \ldots, m; j = 1, \ldots, p)
$$

In the particular case $p = 1$, **C** is an $n \times 1$ vector (**x** say), and the product $\mathbf{A} = \mathbf{Bx}$ is an $m \times 1$ vector (**y** say) with

$$
y_i = a_{i1} \quad \text{and} \quad c_{k1} = x_k
$$

In a BASIC program $a_{ij}$, $b_{ik}$ and $c_{kj}$ may be represented by matrix variables A(I, J), B(I, K), C(K, J). Then A(I, J) may be determined within loops on I (from 1 to M) and on J (from 1 to P), the sum of the products B(I, K)*C(K, J) being accumulated in a loop on K, starting with an initial value of 0. (Compare the simpler Program 3.2 of Reference 1, where the sum of a set of $n$ numbers is determined.) A formal program, Program 4.1, is given below. This illustrates not only the current task, but also the inclusion of DIM, INPUT, and PRINT routines in a complete program.

**Program 4.1** MATMUL: Matrix multiplication

```
LIST
MATMUL

10      REM- MATMUL: CALCULATES MATRIX PRODUCT A*B
20      DIM A(10,10),B(10,10),C(10,10)
30      PRINT "NO OF ROWS AND COLS IN A ";
40      INPUT M,N
50      PRINT "ELEMENTS OF A (1 PER LINE)"
60      FOR I=1 TO M
70      FOR J=1 TO N
80      INPUT A(I,J)
90      NEXT J
100     NEXT I
110     PRINT "NO OF ROWS AND COLS IN B ";
120     INPUT N1,P
130     IF N1=N THEN 160
140     PRINT "A AND B ARE INCOMPATIBLE"
150     GO TO 350
160     PRINT "ELEMENTS OF B (1 PER LINE)"
170     FOR I=1 TO N
180     FOR J=1 TO P
190     INPUT B(I,J)
200     NEXT J
210     NEXT I
220     PRINT "NO OF ROWS AND COLS IN C :";
230     PRINT M;P
240     PRINT "ELEMENTS OF C :"
250     FOR I=1 TO M
260     FOR J=1 TO P
270     D=0
280     FOR K=1 TO N
290     D=D+A(I,K)*B(K,J)
300     NEXT K
310     C(I,J)=D
320     PRINT C(I,J)
330     NEXT J
340     NEXT I
350     END

Ready
```

*Sample run 1*

```
RUN
MATMUL

NO OF ROWS AND COLS IN A ? 1,1
ELEMENTS OF A (1 PER LINE)
? 1
NO OF ROWS AND COLS IN B ? 2,2
A AND B ARE INCOMPATIBLE
Ready
```

*Sample run 2*

```
RUN
MATMUL

NO OF ROWS AND COLS IN A ?  3,2
ELEMENTS OF A (1 PER LINE)
? 3
? 2
? 5
? -1
? 4
? 9
NO OF ROWS AND COLS IN B ?  2,4
ELEMENTS OF B (1 PER LINE)
? 1
? 5
? 7
? 2
? 4
? 8
? 1
? 6
NO OF ROWS AND COLS IN C :  3   4
ELEMENTS OF C :
  11
  31
  23
  18
  1
  17
  34
  4
  40
  92
  37
  62
Ready
```

## 4.4 Reference

1. Mason, J.C., *BASIC Numerical Mathematics*, Butterworths (1983).

## PROBLEMS

**(4.1)** Write BASIC routines, without using MAT statements, to perform the following assignments:

  (i) $\mathbf{A} = \mathbf{B}$    (ii) $\mathbf{A} = \mathbf{B} + \mathbf{C}$    (iii) $\mathbf{A} = \mathbf{B} - \mathbf{C}$
  (iv) $\mathbf{A} = K\mathbf{B}$ ($K$ scalar)        (v) $\mathbf{A} = \mathbf{O}$
  (vi) $\mathbf{A} = m \times n$ matrix of ones.

**(4.2)** Write a complete program to input a matrix $\mathbf{A}$, replace $\mathbf{A}$ by $\mathbf{A}^T$ (in the same variable locations A(I, J)), and output $\mathbf{A}^T$. Note: Another matrix $\mathbf{B}$ must be used.

**(4.3)** Write complete programs to perform tasks (ii) and (vi), respectively, of Problem 1.

Chapter 5

# Matrix calculations: direct methods

## ESSENTIAL THEORY

### 5.1 Introduction

In the previous chapter we discussed the development of BASIC routines for the elementary matrix statements in Table 1.1, and in particular we wrote our own program for the calculation of products of matrices and/or vectors. In the present chapter we focus attention on the most important application of matrices, namely the solution of a well-determined system of $n$ linear equations in $n$ unknowns:

$$\left.\begin{array}{l} a_{11}x_1 + a_{12}x_2 + \ldots + a_{1n}x_n = b_1 \\ a_{21}x_1 + a_{22}x_2 + \ldots + a_{2n}x_n = b_2 \\ \vdots \qquad\qquad\qquad\qquad\quad \vdots \\ a_{n1}x_1 + a_{n2}x_2 + \ldots + a_{nn}x_n = b_n \end{array}\right\} \qquad (5.1)$$

where $a_{11}, \ldots, a_{nn}$ are specified coefficients, $b_1, \ldots, b_n$ are specified right-hand sides, and $x_1, \ldots, x_n$ are required unknowns. As discussed in Chapter 3, (5.1) can be expressed simply in the matrix form

$$\mathbf{Ax} = \mathbf{b} \qquad (5.2)$$

where $\mathbf{A}$ is an $n \times n$ matrix of coefficients $a_{ij}$, $\mathbf{b}$ is an $n \times 1$ vector of right-hand sides and $\mathbf{x}$ is an $n \times 1$ vector of unknowns. If $\mathbf{A}$ is non-singular, that is to say if the determinant $|\mathbf{A}|$ of $\mathbf{A}$ is non-zero, then $\mathbf{A}^{-1}$ is well-defined and (5.2) has the unique solution

$$\mathbf{x} = \mathbf{A}^{-1}\mathbf{b} \qquad (5.3)$$

In discussing the solution of (5.2), we shall also cover a number of related problems, including

(i) the evaluation of $|\mathbf{A}|$,
(ii) the determination of the matrix $\mathbf{A}^{-1}$,
(iii) the determination (Problem 7) of the unknown solution $\mathbf{X}$ of the general matrix equation $\mathbf{AX} = \mathbf{B}$.

40

## 5.2 BASIC matrix routines

If BASIC matrix routines are available (see Table 1.1), then a very simple (though computationally rather inefficient) program may be written to exploit these routines for the problem (5.2). Specifically the solution $\mathbf{x}$ of (5.2) may be obtained from the formula (5.3), by first determining the matrix $\mathbf{C} = \mathbf{A}^{-1}$ by the instruction

MAT C = INV(A)

and hence calculating $\mathbf{x} = \mathbf{A}^{-1}\mathbf{b} = \mathbf{C}\mathbf{b}$ by the instruction

MAT X = C*B

Having determined $\mathbf{x}$, the residual vector

$$\mathbf{r} = \mathbf{b} - \mathbf{A}\mathbf{x} \tag{5.4}$$

may be calculated to provide a check on accuracy. This residual vector is not necessarily a vector of zeros, on account of rounding errors, and the sizes of the elements of $\mathbf{r}$ measure the accuracy with which equations (5.1) have been solved.

Program 5.1 inputs a matrix $\mathbf{A}$ (row by row) and a vector $\mathbf{b}$, using matrix INPUT statements. It then calculates

$$\mathbf{A}^{-1}, \quad \mathbf{x} = \mathbf{A}^{-1}\mathbf{b}, \quad \mathbf{d} = \mathbf{A}\mathbf{x}, \text{ and } \mathbf{r} = \mathbf{b} - \mathbf{A}\mathbf{x}$$

by using appropriate BASIC matrix routines. Clearly $\mathbf{d}$ should prove to be approximately equal to $\mathbf{b}$, and $\mathbf{r}$ should prove to be approximately zero. The program tests the case

$$\mathbf{A} = \begin{bmatrix} 1 & 1 & 1 \\ 1 & 2 & 5 \\ 1 & 5 & 10 \end{bmatrix}, \mathbf{b} = \begin{bmatrix} 3 \\ 8 \\ 16 \end{bmatrix}$$

for which the true solution is

$$\mathbf{x} = \begin{bmatrix} 1 \\ 1 \\ 1 \end{bmatrix}$$

**Program 5.1** MATINV: Solution of equations by BASIC matrix routines

```
LIST
MATINV

10   REM- MATINV: SOLVES AX=B USING BASIC MATRIX ROUTINES
20   DIM A(20,20),C(20,20),X(20,1),B(20,1),R(20,1)
30   PRINT "NUMBER OF EQUATIONS";
40   INPUT N
```

```
50   PRINT "INPUT MATRIX A ROW BY ROW"
60   MAT INPUT A(N,N)
70   PRINT "INPUT VECTOR B "
80   MAT INPUT B(N,1)
90   MAT C=ZER(N,N)
100  MAT X=ZER(N,1)
110  MAT R=ZER(N,1)
120  MAT C=INV(A)
130  MAT X=C*B
140  MAT R=A*X
150  PRINT "INVERSE MATRIX:"
160  MAT PRINT C
170  PRINT "SOLUTION :"
180  MAT PRINT X
190  PRINT "PRODUCT A*X :"
200  MAT PRINT R
210  MAT R=B-R
220  PRINT "RESIDUAL B-A*X :"
230  MAT PRINT R
240  END

Ready
```

## Sample run

```
RUN
MATINV

NUMBER OF EQUATIONS? 3
INPUT MATRIX A ROW BY ROW
? 1,1,1,1,2,5,1,5,10
INPUT VECTOR B
? 3,8,16
INVERSE MATRIX:
 .714286
 .714286
-.428571
 .714286
-1.28571
 .571429
-.428571
 .571429
-.142857
SOLUTION :
 1
 1
 1
PRODUCT A*X :
 3
 8
 16
RESIDUAL B-A*X :
-.238419E-06
-.190735E-05
-.38147E-05
Ready
```

## Program notes
(1) Dynamic dimensioning, which is not available on all implementations of BASIC, has been avoided. Instead, it has been achieved *implicitly* in the input statements 60 and 80. For example the statement

60  MAT INPUT A(N, N)

will automatically redimension A from 20 by 20 to N by N. Similarly the statements 90, 100, 110 implicitly redimension C, X and R by equating them to zero matrices of the required dimensions. For example X is redimensioned from 20 by 1 to N by 1.

(2) On computers which *do* have dynamic dimensioning, the following changes to the program could be made:

DELETE 20
45 DIM A(N, N), C(N, N), X(N, 1), B(N, 1), R(N, 1)
DELETE 90–110

(3) The vectors **x** and **r** have been represented here as rectangular matrices with *n* rows and 1 column, using 2 subscripts. The reason for this is that, in the BASIC implementation used here, the matrix product **AX** can only be performed for matrices **A** and **X** with 2 subscripts specified. In some computers it may be possible to specify only one subscript for a vector, in which case suitable simplifications may be made. But be careful!

### 5.3 Gauss elimination: simple approach

Although Program 5.1 is very convenient to use, it has two particular disadvantages:

  (i) It is inefficient, since it calculates **x** by first calculating $\mathbf{A}^{-1}$. Indeed it involves about four times as many calculations as are really necessary.

 (ii) It uses a 'black box' to solve the problem, namely the built-in BASIC matrix routine to calculate $\mathbf{A}^{-1}$, and this cannot easily be understood or developed.

It is therefore desirable for us to write a computer program which we can understand, have access to, and extend to provide added features. As pointed out in Chapter 2, the method to be used is not Cramer's rule which involves about $(n + 1)!$ multiplications, but the 'Gauss Elimination' method which involves about $\frac{1}{3}n^3$ multiplications. The main idea of this method is to simplify the structure of the equations by adding together suitable multiples of them.

### 5.3.1 *Augmented matrix*

If multiples of individual equations in (5.1) are added together, then this has the effect of adding together corresponding rows of the matrix **A** and vector **b**. Indeed there is an exact (one-to-one) correspondence between the system (5.1) and the 'augmented matrix'

$$(\mathbf{A}|\mathbf{b})$$

consisting of the elements of **A** and **b** placed side by side to form an *n* by $(n + 1)$ matrix. The 'partition' drawn between **A** and **b** is purely for clarification and is not essential to the logic. We may thus add multiples of the rows of the augmented matrix together as often as we wish, and then reinterpret the resulting matrix as corresponding to a new system of equations equivalent to (5.1).

For example, the system

$$\left.\begin{array}{r} x_1 + 0.5\ x_2 + 0.33x_3 = 1.8 \\ 0.5\ x_1 + 0.33x_2 + 0.25x_3 = 1.1 \\ 0.33x_1 + 0.25x_2 + 0.2\ x_3 = 0.78 \end{array}\right\} \qquad (5.5)$$

may be represented by the augmented matrix

$$\left[\begin{array}{ccc|c} 1 & 0.5 & 0.33 & 1.8 \\ 0.5 & 0.33 & 0.25 & 1.1 \\ 0.33 & 0.25 & 0.2 & 0.78 \end{array}\right] \qquad (5.6)$$

### 5.3.2 Gauss elimination: exact

Suppose the notation

$$R_2 \to R_2 + mR_1 \qquad (5.7)$$

is used to mean 'add the multiple $m$ of row one to row two in the given augmented matrix'. Then the row operations

$$R_2 \to R_2 - 0.5R_1, \ R_3 \to R_3 - 0.33R_1$$

change the augmented matrix to

$$\left[\begin{array}{ccc|c} 1 & 0.5 & 0.33 & 1.8 \\ 0 & 0.08 & 0.085 & 0.2 \\ 0 & 0.085 & 0.0911 & 0.186 \end{array}\right]$$

and reduce to zero the elements below the diagonal in the first column of **A**. Any element which is used to eliminate others in the same column (such as in this case the element 1 in the first row and first column) is called a 'pivot' and will be highlighted in bold type, and any constants $m$ used in the row operations (such as in this case $-0.5$ and $-0.33$) are called 'multipliers'.

The further row operation

$$R_3 \to R_3 - \frac{0.085}{0.08}R_2 = R_3 - 1.0625R_2$$

changes the augmented matrix to

$$\left[\begin{array}{ccc|c} 1 & 0.5 & 0.33 & 1.8 \\ 0 & 0.08 & 0.085 & 0.2 \\ 0 & 0 & 0.0007875 & -0.0265 \end{array}\right] \qquad (5.8)$$

and reduces to zero the element(s) below the diagonal in the second column of **A**. Here the pivot is 0.08 in the second row and second column, and the multiplier is $-1.0625$. We see that the matrix **A** (to the left of the partition) has now been reduced to an 'upper triangular' matrix with no non-zero entries below its diagonal. The augmented matrix may now be translated back into the system of equations

$$\left.\begin{array}{rlrl} x_1 + 0.5 \ x_2 + 0.33 & & x_3 = & 1.8 \\ 0.08x_2 + 0.085 & & x_3 = & 0.2 \\ 0.0007875 \ x_3 = & -0.0265 \end{array}\right\} \qquad (5.9)$$

These equations may obviously be solved in reverse order by 'back-substitution' to give

$$\left.\begin{array}{rll} x_3 = -0.0265/0.0007875 & = & -33.651 \\ x_2 = (0.2 - 0.085x_3)/0.08 & = & 38.254 \\ x_1 = 1.8 - 0.5x_2 - 0.33x_3 = & & -6.222 \end{array}\right\} \qquad (5.10)$$

The above process in which row operations are performed, with pivots taken on the diagonal of **A** to eliminate all elements below the diagonal, is called 'Gauss elimination'.

It is worth remarking that the particular equations (5.5) are very ill-conditioned. Indeed if the first two right-hand sides are changed by about two per cent from 1.8 and 1.1 to 1.83 and 1.08, respectively, then it is easily seen that the exact solution of the equations becomes

$$x_1 = x_2 = x_3 = 1$$

This solution bears no resemblance whatsoever to the solution (5.10) of (5.5)! Thus any errors in the data of (5.5) are bound to have a considerable effect on the solution, even when this is calculated exactly from the given data.

### 5.3.3  *Gauss elimination: inexact*

In practice Gauss elimination is not normally performed exactly and so, in addition to possible errors in the data, account must be taken of errors in the method. Although truncation errors are completely absent, since the method has a finite number of steps, *rounding errors* are present and may have a strong effect.

For example, if the above exact calculations for (5.5) were instead performed inexactly, with just two significant figures retained after every arithmetic operation (and with fives rounded up), the method would progress as follows:

$$\begin{bmatrix} 1 & 0.5 & 0.33 \\ 0.5 & 0.33 & 0.25 \\ 0.33 & 0.25 & 0.2 \end{bmatrix} \begin{array}{|c} 1.8 \\ 1.1 \\ 0.78 \end{array} \rightarrow \begin{bmatrix} 1 & 0.5 & 0.33 \\ 0 & 0.08 & 0.08 \\ 0 & 0.08 & 0.09 \end{bmatrix} \begin{array}{|c} 1.8 \\ 0.2 \\ 0.19 \end{array}$$

$$\rightarrow \begin{bmatrix} 1 & 0.5 & 0.33 \\ 0 & 0.08 & 0.08 \\ 0 & 0 & 0.01 \end{bmatrix} \begin{array}{|c} 1.8 \\ 0.2 \\ -0.01 \end{array}$$

with multipliers

$$m_{21} = -0.5, \ m_{31} = -0.33, \text{ and } m_{32} = -1$$

Here each multiplier is given a name $m_{ij}$, the two suffices $i$ and $j$ corresponding precisely to the row and column numbers of the element $a_{ij}$ of **A** which is eliminated. Back substitution from the final augmented matrix now gives

$$x_3 = -1, \quad x_2 = 3.5, \quad x_1 = 0.33 \tag{5.11}$$

and once again the solution bears no resemblance whatever to the exact solution (5.10).

Nevertheless, if we substitute (5.11) into (5.5) and use exact arithmetic, we find that

$$\mathbf{Ax} = \begin{bmatrix} 1.75 \\ 1.07 \\ 0.784 \end{bmatrix}$$

and

$$\mathbf{r} = \mathbf{b} - \mathbf{Ax} = \begin{bmatrix} 0.05 \\ 0.03 \\ -0.004 \end{bmatrix}$$

Thus **Ax** is close to the vector **b**, and the residual **r** is small. In summary, the solution (5.11) is acceptable if we are concerned only with satisfying the equations (5,5) accurately, but unacceptable if we are concerned with obtaining accurate values of each component of **x**.

Although the above discussion of two decimal place arithmetic may seem unrealistic in the context of BASIC programming, which uses six to nine decimals, it does in fact provide a good model for the sort of difficulties which are encountered in BASIC when a large number of equations are solved. Indeed as few as ten equations can sometimes present severe difficulties in BASIC.

### 5.3.4 Calculation of the determinant of **A**

Row operations in Gauss elimination, such as (5.7), do not affect the values of the determinant of **A**. Indeed one of the traditional ways of calculating a determinant involves simplifying the entries by adding multiples of one row to another. So, for example, the final 'upper triangular matrix' to the left of the partition in (5.8) has the same determinant as the original matrix **A** in (5.6), and hence the determinant is easily calculated as the product of the diagonal terms in (5.8). Specifically

$$|\mathbf{A}| = \begin{vmatrix} 1 & 0.5 & 0.33 \\ 0.5 & 0.33 & 0.25 \\ 0.33 & 0.25 & 0.2 \end{vmatrix} = \begin{vmatrix} 1 & 0.5 & 0.33 \\ 0 & 0.08 & 0.085 \\ 0 & 0 & 0.0007875 \end{vmatrix}$$

$$= (1)(0.08)(0.0007875) = 0.000063$$

Notice that in this example the determinant is very small, which indicates that **A** is close to being *singular*. This means that each of the three left-hand sides in (5.5) is nearly a linear combination of the other two, and hence the three equations are close to being either inconsistent with each other (and hence having no solution) or reducible to just two equations (and hence having many solutions).

In a physical problem, such a phenomenon could have important practical implications. It almost invariably implies an *ill-conditioned* problem.

### 5.3.5 A short Gauss elimination program

We shall now start developing algorithms and programs for performing Gauss elimination to solve a general set of $n$ equations in $n$ unknowns. Some of the programs will also calculate the residual vector $\mathbf{r} = \mathbf{b} - \mathbf{A}\mathbf{x}$ (to test the accuracy of the solution) and the determinant $|\mathbf{A}|$ (to test the singularity of **A** and the conditioning of the problem).

The key step in writing a program is to decide what happens to the matrix **A** and right-hand side **b** at any intermediate cycle of the elimination process. In cycle $(k)$, say, where $k$ is any number from 1 to $n - 1$, elements below the diagonal in the $k$th column of **A** are eliminated. Before this cycle, the augmented matrix has the form

$$\left[ \begin{array}{ccccccc|c} a_{11}^{(k)} & a_{12}^{(k)} & \cdots & & \cdots & a_{1n}^{(k)} & & b_1^{(k)} \\ & a_{22}^{(k)} & \cdots & & \cdots & a_{2n}^{(k)} & & b_2^{(k)} \\ & & \ddots & & & \vdots & & \vdots \\ \mathbf{O} & & & a_{kk}^{(k)} & \cdots & a_{kn}^{(k)} & & b_k^{(k)} \\ & & & a_{k+1,k}^{(k)} & \cdots & a_{k+1,n}^{(k)} & & b_{k+1}^{(k)} \\ & & & \vdots & & \vdots & & \vdots \\ & & & a_{n,k}^{(k)} & \cdots & a_{n,n}^{(k)} & & b_n^{(k)} \end{array} \right] \quad (5.12)$$

where the large zero indicates that all elements in this portion of the matrix are zeros. The superscript $(k)$ is used to indicate that the elements $a_{11}, \ldots, a_{nn}, b_1, \ldots, b_n$ of the augmented matrix have the values that have been assigned to them before cycle $(k)$. Similarly $\mathbf{A}^{(k)}$ and $\mathbf{b}^{(k)}$ denote the matrices on either side of the vertical partition in (5.12). Cycle $(k)$ proceeds as follows, with $a_{kk}^{(k)}$ as pivot:

(i) Calculate multipliers to eliminate the element $a_{ik}^{(k)}$ in row $i$ and column $k$ (for $i = k + 1, \ldots, n$):

$$m_{ik} = -a_{ik}^{(k)}/a_{kk}^{(k)} \tag{5.13}$$

(ii) Add a multiple $m_{ik}$ of row $k$ of $\mathbf{A}^{(k)}$ to row $i$ (for $i = k + 1, \ldots, n$):

$$a_{ij}^{(k+1)} = a_{ij}^{(k)} + m_{ik}a_{kj}^{(k)} \quad j = k, \ldots, n \tag{5.14}$$

(iii) Add a multiple $m_{ik}$ of row $k$ of $\mathbf{b}^{(k)}$ to row $i$ (for $i = k + 1, \ldots, n$):

$$b_i^{(k+1)} = b_i^{(k)} + m_{ik}b_k^{(k)} \tag{5.15}$$

All other elements of $\mathbf{A}$ and $\mathbf{b}$ are left unchanged, so that:

$$\left. \begin{array}{l} a_{ij}^{(k+1)} = a_{ij}^{(k)} \text{ for } i = 1, \ldots, n; j = 1, \ldots, k-1 \\ \qquad \qquad \text{and } i = 1, \ldots, k; j = 1, \ldots, n \\ b_i^{(k+1)} = b_i^{(k)} \text{ for } i = 1, \ldots, k. \end{array} \right\} \tag{5.16}$$

The augmented matrix now has the form of (5.12), but with $k$ replaced by $k + 1$, and so we can move on to cycle $(k + 1)$. After $n - 1$ cycles $(k = 1, 2, \ldots, n - 1)$ have been completed, the augmented matrix reduces to

$$\left[ \begin{array}{cccc|c} a_{11}^{(n)} & a_{12}^{(n)} & \cdots & a_{1n}^{(n)} & b_1^{(n)} \\ & a_{22}^{(n)} & \cdots & a_{2n}^{(n)} & b_2^{(n)} \\ & & \ddots & \vdots & \vdots \\ \mathbf{O} & & & a_{nn}^{(n)} & b_n^{(n)} \end{array} \right] \tag{5.17}$$

The augmented matrix now corresponds to the set of linear equations

$$\begin{array}{l} a_{11}^{(n)}x_1 + a_{12}^{(n)}x_2 + \ldots + a_{1n}^{(n)}x_n = b_1^{(n)} \\ \qquad a_{22}^{(n)}x_2 + \ldots + a_{2n}^{(n)}x_n = b_2^{(n)} \\ \qquad \qquad \qquad \vdots \\ \qquad \qquad \qquad a_{nn}^{(n)}x_n = b_n^{(n)} \end{array} \tag{5.18}$$

which may be solved in the reverse order by back substitution for $x_n, x_{n-1}, \ldots, x_1$ as follows:

(iv) $x_n = b_n^{(n)}/a_{nn}^{(n)}$

$$x_i = (b_i^{(n)} - \sum_{j=i+1}^{n} a_{ij}^{(n)} x_j)/a_{ii}^{(n)} \qquad\qquad (5.19)$$

$$(i = n-1, \ldots, 1)$$

Observe that in programming the mathematics, the superscripts $(k)$ are not required. We may simply replace equations (5.14) and (5.15) by

$$a_{ij} := a_{ij} + m_{ik}a_{kj}, \quad b_i := b_i + m_{ik}b_k$$

the elements $a_{ij}$ and $b_i$ being updated at each cycle. Equations (5.16) are then unnecessary, as also is the case $j = k$ in (5.14), and steps (i), (ii), (iii), (iv) give the following algorithm.

*Algorithm 5.2 Gauss elimination without interchanges, short version*
Elimination Process:
For $k = 1, 2, \ldots, n-1$:
For $i = k+1, \ldots, n$:
  (i) $m_{ik} = -a_{ik}/a_{kk}$
  (ii) $a_{ij} := a_{ij} + m_{ik}a_{kj}$ $(j = k+1, \ldots, n)$
  (iii) $b_i := b_i + m_{ik}b_k$
Back substitution:
(iv) $x_n = b_n/a_{nn}$
For $i = n-1, \ldots, 1$:

$$x_i = (b_i - \sum_{j=i+1}^{n} a_{ij}x_j)/a_{ii}$$

Algorithm 5.2 is coded in BASIC as Program 5.2, which follows. This program is given the title GSHORT, to indicate that it is a short program illustrating the core of a Gauss elimination method. It is, however, not a foolproof program, as the Program Notes explain. Better, but inevitably longer, versions of Gauss elimination will therefore follow.

**Program 5.2** GSHORT: Gauss elimination without interchanges, short version

```
LIST
GSHORT

10   REM- GSHORT: SOLVES AX=B BY GAUSS ELIMINATION (NO INTERCHANGES)
30   DIM A(20,20),X(20),B(20)
50   PRINT "NUMBER OF EQUATIONS ";
60   INPUT N
70   PRINT "INPUT MATRIX A ROW BY ROW "
80   MAT INPUT A(N,N)
90   PRINT "INPUT VECTOR B"
100  MAT INPUT B(N)
140  REM- PERFORMS N-1 CYCLES OF ELIMINATIONS
170  REM- CURRENT CYCLE NUMBER IS K
```

```
180 FOR K=1 TO N-1
190 REM- CALCULATES MULTIPLIER M(I,K) TO ELIMINATE A(I,K)
200 FOR I=K+1 TO N
210 M=-A(I,K)/A(K,K)
220 REM- PERFORMS ROW OPERATIONS ON A
230 FOR J=K+1 TO N
240 A(I,J)=A(I,J)+M*A(K,J)
250 NEXT J
260 REM- PERFORMS ROW OPERATIONS ON B
270 B(I)=B(I)+M*B(K)
280 NEXT I
290 NEXT K
360 REM- CALCULATES X(N),...,X(1) BY BACK SUBSTITUTION
370 X(N)=B(N)/A(N,N)
380 FOR I=N-1 TO 1 STEP -1
390 D=B(I)
400 FOR J=I+1 TO N
410 D=D-A(I,J)*X(J)
420 NEXT J
430 X(I)=D/A(I,I)
440 NEXT I
450 PRINT "SOLUTION X :"
460 MAT PRINT X(N)
590 END

Ready
```

## Sample run 1

```
RUN
GSHORT

NUMBER OF EQUATIONS ? 3
INPUT MATRIX A ROW BY ROW
? 1,1,1,1,2,5,1,5,10
INPUT VECTOR B
? 3,8,16
SOLUTION X :
 1
 1
 1
Ready
```

## Sample run 2

```
RUN
GSHORT

NUMBER OF EQUATIONS ? 2
INPUT MATRIX A ROW BY ROW
? 1,1,1,1
INPUT VECTOR B
? 2,2
%BAS-F-DIVBY_ZER, Division by 0
-BAS-I-USEPC_PSL,  at user PC=00116AC1, PSL=03C0002A
-BAS-I-FROLINMOD,  from line 370 in module GSHORT
--SYSTEM-F-FLTDIV, arithmetic trap, floating/decimal divide by zero
Ready
```

## Sample run 3

```
RUN
GSHORT

NUMBER OF EQUATIONS ? 2
INPUT MATRIX A ROW BY ROW
? 0,1,1,1
INPUT VECTOR B
? 1,2
%BAS-F-DIVBY_ZER, Division by 0
-BAS-I-USEPC_PSL,  at user PC=00116999, PSL=03C0002A
-BAS-I-FROLINMOD,  from line 210 in module GSHORT
--SYSTEM-F-FLTDIV, arithmetic trap, floating/decimal divide by zero
Ready
```

*Sample run 4*

```
RUN
GSHORT

NUMBER OF EQUATIONS ? 2
INPUT MATRIX A ROW BY ROW
? .000001,1,1,1
INPUT VECTOR B
? 1.000001,2
SOLUTION X :
 1.01328
 1
Ready
```

*Program notes*

(1) The program is a direct translation of Algorithm 5.2 (which becomes even more apparent if all REM statements are deleted). The only nontrivial coding is that required in part (iv) of the algorithm, which involves a summation $\sum$ (compare Program 3.2 of Reference 1). The $\sum$ coding is carried out in lines 390–420, where $-a_{ij}x_j$ is summed over $j$ and added to $b_i$, the result being accumulated in the location D.

(2) Matrix input and print routines have been used (for brevity in the program and data) in instructions 80, 100, 460. For computers without matrix routines the following changes may be made.

```
 80 FOR I = 1 TO N
 81 FOR J = 1 TO N
 82 INPUT A(I, J)
 83 NEXT J
 84 NEXT I
100 FOR I = 1 TO N
101 INPUT B(I)
102 NEXT I
460 FOR I = 1 TO N
461 PRINT X(I)
462 NEXT I
```

Note that, if these changes are made, then the data must be input one number per line.

(3) The program will of course fail if $|\mathbf{A}| = 0$, since there is then no unique solution (see Sample run 2).

(4) The program will fail if, at any stage of the elimination process, one of the diagonal elements of $\mathbf{A}$ is zero. This will lead to a division by zero, either in the calculation of a multiplier or in the back substitution calculation (see Sample run 3).

(5) The program will fail if, at any stage, one of the diagonal elements of $\mathbf{A}$ is 'very close' to zero. For this may result in 'accumulator

overflow', i.e. a number with an exponent too large for the computer to handle.

(6) In certain cases the program may actually produce a 'wrong' answer. Indeed in Sample run 4 the correct solution of $x_1 = x_2 = 1$ is only obtained within 1 per cent, and moreover the linear equations are not satisfied.

The Program notes 4 to 6 may seem rather discouraging. However, we have so far only implemented (for simplicity and comprehension) the 'raw' Gauss elimination algorithm, and clearly changes should be made. Program note 6 is particularly disturbing, since there is at present nothing in the computer output to indicate a poor result. The first improvement that we must therefore make is to include a check on the accuracy of the claimed 'solution'. This is easily done, and Program 5.3 below includes the calculation of the residual vector

$$\mathbf{r} = \mathbf{b} - \mathbf{A}\mathbf{x}$$

where $\mathbf{A}$ and $\mathbf{b}$ are the original matrix and right-hand side.

### 5.3.6 *An informative Gauss elimination program without interchanges*

If the residual $\mathbf{r}$ has small components, then the system of equations $\mathbf{A}\mathbf{x} = \mathbf{b}$ is well satisfied, and this may be all that we require for our solution $\mathbf{x}$. However, if accurate components of $\mathbf{x}$ are also required, then success will depend on the conditioning of $\mathbf{A}$. The matrix is likely to be ill-conditioned if $|\mathbf{A}|$ is 'small', and so the calculation of $|\mathbf{A}|$ is relevant. Its value is cheaply determined during Gauss elimination, as was pointed out in Section 5.3.4, and indeed

$$|\mathbf{A}| = a_{11}^{(n)} a_{22}^{(n)} \ldots a_{nn}^{(n)}$$

the product of the diagonal terms of $\mathbf{A}$ after the completion of the elimination process.

Algorithm 5.3 and Program 5.3 incorporate these calculations. Program 5.3 is given the name GAUSS, and it is an 'elementary' Gauss elimination program.

### Algorithm 5.3 *Gauss elimination without interchanges, standard version*

This is identical to Algorithm 5.2, apart from the following addenda:

(v) Calculate $|\mathbf{A}| = a_{11} a_{22} \ldots a_{nn}$ after row operations on $\mathbf{A}$.

(vi) Calculate $\mathbf{r} = \mathbf{b} - \mathbf{A}\mathbf{x}$
   (where $\mathbf{A}$ and $\mathbf{b}$ are the original matrix and right-hand side (before row operations).

If each element of **r** is small (compared with elements of **A**) then the equations **Ax** = **b** have been satisfactorily solved.

If |**A**| is small, then the equations are probably ill-conditioned and the vector **x** may be inaccurate.

**Program 5.3** GAUSS: Gauss elimination without interchanges, standard version

```
LIST
GAUSS

10   REM- GAUSS: SOLVES AX=B BY GAUSS ELIMINATION (NO INTERCHANGES)
20   REM- ALSO CALCULATES RESIDUAL VECTOR B-AX AND DETERMINANT OF A
30   DIM A(20,20),X(20),B(20)
40   DIM A1(20,20),B1(20),R(20)
50   PRINT "NUMBER OF EQUATIONS ";
60   INPUT N
70   PRINT "INPUT MATRIX A ROW BY ROW."
80   MAT INPUT A(N,N)
90   PRINT "INPUT VECTOR B"
100  MAT INPUT B(N)
110  REM- KEEPS COPIES OF ORIGINAL A,B IN A1,B1
120  MAT A1=A
130  MAT B1=B
140  REM- PERFORMS N-1 CYCLES OF ELIMINATIONS
150  REM- ACCUMULATES DETERMINANT OF A IN D1
160  D1=1
170  REM- CURRENT CYCLE NUMBER IS K
180  FOR K=1 TO N-1
190  REM- CALCULATES MULTIPLIER M(I,K) TO ELIMINATE A(I,K)
200  FOR I=K+1 TO N
210  M=-A(I,K)/A(K,K)
220  REM- PERFORMS ROW OPERATIONS ON A
230  FOR J=K+1 TO N
240  A(I,J)=A(I,J)+M*A(K,J)
250  NEXT J
260  REM- PERFORMS ROW OPERATIONS ON B
270  B(I)=B(I)+M*B(K)
280  NEXT I
290  NEXT K
300  REM- CALCULATES DETERMINANT D1 OF A
310  FOR I=1 TO N
320  D1=D1*A(I,I)
330  NEXT I
340  PRINT "DETERMINANT OF A :";
350  PRINT D1
360  REM- CALCULATES X(N),...,X(1) BY BACK SUBSTITUTION
370  X(N)=B(N)/A(N,N)
380  FOR I=N-1 TO 1 STEP -1
390  D=B(I)
400  FOR J=I+1 TO N
410  D=D-A(I,J)*X(J)
420  NEXT J
430  X(I)=D/A(I,I)
440  NEXT I
450  PRINT "SOLUTION X :"
460  MAT PRINT X(N)
470  REM- CALCULATES RESIDUAL VECTOR R=B-AX
480  FOR I=1 TO N
490  W=B1(I)
500  FOR J=1 TO N
510  W=W-A1(I,J)*X(J)
520  NEXT J
530  R(I)=W
540  NEXT I
550  PRINT "RESIDUAL VECTOR B-AX :"
560  MAT PRINT R(N)
570  END

Ready
```

*Sample run*

```
RUN
GAUSS

NUMBER OF EQUATIONS ? 2
INPUT MATRIX A ROW BY ROW
? .000001,1,1,1
INPUT VECTOR B
? 1.000001,2
DETERMINANT OF A :-.999999
SOLUTION X :
 1.01328
 1
RESIDUAL VECTOR B-AX :
 0
-.132789E-01
Ready
```

*Program notes*

(1) The program includes *all* numbered statements that appear in Program 5.2, plus the additional statements numbered 20, 40, 110–130, 150, 160, 300–350, 470–560.

(2) The original elements A(I, J) and B(I) of **A** and **b** are stored in locations A1(I, J) and B1(I) for use in the calculation of the residual vector in statements 470–560.

(3) Matrix routines are used to equate A1(I, J) to A(I, J) and B1(I) to B(I) in statements 120–130. To avoid such matrix routines, write:

```
120  FOR I = 1 TO N
121  FOR J = 1 TO N
122  A1(I, J) = A(I, J)
123  NEXT J
124  B1(I) = B(I)
130  NEXT I
```

(4) The symbol D1 is used for the determinant, and this is calculated in statements 160, 310–330. The routine used is identical to that adopted in Program 3.1 of Reference 1 for a product of a set of numbers.

(5) All Program notes of Program 5.2 still hold true!

Clearly this algorithm has to be modified further if good results are to be obtained consistently, and indeed in the sample run a large residual vector is still obtained. However, at least Program 5.3 can be used with some confidence in the sense that we now know whether or not the equations have been satisfied.

Before giving the next improved algorithm (Section 5.4), let us count the number of arithmetic operations and verify that it is approximately $n^3/3$.

### 5.3.7 *Operations count*

The number of arithmetic operations of addition, multiplication, etc. involved in Gauss elimination may easily be deduced by simply counting the number of these operations in parts (i), (ii), (iii), and (iv) of Algorithm 5.2, and taking account of the loops $k = 1, \ldots,$ $n - 1; i = k + 1, \ldots, n; j = k + 1, \ldots, n$, etc. For simplicity we only count multiplications and equate a division to a multiplication.

(i) Number of multiplications $= \sum\limits_{k=1}^{n-1} \sum\limits_{i=k+1}^{n} (1) = (n - 1) + (n - 2)$
$$+ \ldots + 1 = \tfrac{1}{2}n(n - 1) \simeq \tfrac{1}{2}n^2$$

(ii) Number $= \sum\limits_{k=1}^{n-1} \sum\limits_{i=k+1}^{n} \sum\limits_{j=k+1}^{n} (1) = (n - 1)^2 + (n - 2)^2 + \ldots + 1^2$
$$= \tfrac{1}{6}(n - 1)n(2n - 1) \simeq \tfrac{1}{3}n^3 - \tfrac{1}{2}n^2$$

(iii) Number $=$ same as in (i) $\simeq \tfrac{1}{2}n^2$

(iv) Number $= 1 + \sum\limits_{i=1}^{n-1} \left( 1 + \sum\limits_{j=i+1}^{n} (1) \right) = 1 + (n + (n - 1) + \ldots + 2)$
$$= \tfrac{1}{2}n(n + 1) \simeq \tfrac{1}{2}n^2$$

Therefore, the total number of multiplications $\simeq \tfrac{1}{3}n^3 + n^2 \simeq \tfrac{1}{3}n^3$ (for $n$ large).

## 5.4 Gauss elimination: improved approach

Program 5.3 is a cheap and simple approach but (see Program notes) it fails completely if a pivot $a_{kk}$ is zero. Moreover there is one particular source of rounding error in the Gauss elimination method, as implemented in Program 5.3, which can sometimes give large residuals (as we saw in the Sample run) and which emanates from a *practical flaw* in the method. Fortunately this particular flaw can be eradicated by introducing a modification, which also avoids zero pivots. Consider the method applied to the equations

$$0.00100x_1 + 1.00x_2 = 1.00 \qquad (5.20)$$
$$1.00x_1 + 1.00x_2 = 2.00$$

and suppose that three significant figures are used throughout the arithmetic. Then the method progresses as follows:

$$\begin{bmatrix} \mathbf{0.00100} & 1.00 \\ 1.00 & 1.00 \end{bmatrix} \begin{array}{|c} 1.00 \\ 2.00 \end{array} \rightarrow \begin{bmatrix} 0.0010 & 1.00 \\ 0 & -1000 \end{bmatrix} \begin{array}{|c} 1.00 \\ -1000 \end{array}$$

$$(m_{21} = -1000)$$

Back substitution gives

$$x_2 = 1.00, \quad x_1 = 0.00$$

and hence

$$(\mathbf{Ax}) = \begin{bmatrix} 1.00 \\ 1.00 \end{bmatrix} \text{ and } \mathbf{r} = \mathbf{b} - \mathbf{Ax} = \begin{bmatrix} 0 \\ 1.00 \end{bmatrix}$$

In this case $\mathbf{Ax}$ does not approximate $\mathbf{b}$, and $\mathbf{r}$ is far too large to be acceptable.

The problem here is that the large multiplier $m_{21}$ exaggerates the effect of the first equation in (5.20) and effectively annihilates the second equation. Although this is an extreme example, it is easily dealt with. Indeed in general rounding error can be kept under control if all multipliers are made to be less than unity in magnitude. This may be achieved by using as pivot in a particular column not necessarily a diagonal element of $\mathbf{A}$, but rather that entry on or below the diagonal which has largest magnitude. Such a pivot will of course always be non-zero if the matrix is non-singular. To simplify the algebra, we simply interchange the row containing the pivot with the row containing the diagonal element in the augmented matrix, and the algorithm then proceeds as before. This is illustrated for (5.20) as follows:

$$\begin{bmatrix} 0.00100 & 1.00 & \bigm| & 1.00 \\ 1.00 & 1.00 & \bigm| & 2.00 \end{bmatrix} \rightarrow \begin{bmatrix} \mathbf{1.00} & 1.00 & \bigm| & 2.00 \\ 0.00100 & 1.00 & \bigm| & 1.00 \end{bmatrix}$$

$$R_1 \leftrightarrow R_2 \qquad\qquad m_{21} = -0.001$$

$$\rightarrow \begin{bmatrix} 1.00 & 1.00 & \bigm| & 2.00 \\ 0 & 1.00 & \bigm| & 1.00 \end{bmatrix}$$

Here back substitution gives the correct answer $x_1 = x_2 = 1.00$.

A 3 by 3 example using two figure arithmetic provides another illustration of row interchanges (though strictly speaking interchanges are not necessary for accuracy in this case):

$$\begin{bmatrix} 1 & 2 & 3 & \bigm| & 6 \\ 2 & 3 & 4 & \bigm| & 9 \\ 3 & 4 & 6 & \bigm| & 13 \end{bmatrix} \rightarrow \begin{bmatrix} 3 & 4 & 6 & \bigm| & 13 \\ 2 & 3 & 4 & \bigm| & 9 \\ 1 & 2 & 3 & \bigm| & 6 \end{bmatrix} \rightarrow \begin{bmatrix} 3 & 4 & 6 & \bigm| & 13 \\ 0 & 0.3 & 0 & \bigm| & 0.3 \\ 0 & 0.7 & 1.0 & \bigm| & 1.7 \end{bmatrix} \rightarrow$$

$$R_1 \leftrightarrow R_3 \qquad\qquad m_{21} = -0.67 \qquad\qquad R_2 \leftrightarrow R_3$$
$$m_{31} = -0.33$$

$$\begin{bmatrix} 3 & 4 & 6 & \bigm| & 13 \\ 0 & \mathbf{0.7} & 1.0 & \bigm| & 1.7 \\ 0 & 0.3 & 0 & \bigm| & 0.3 \end{bmatrix} \rightarrow \begin{bmatrix} 3 & 4 & 6 & \bigm| & 13 \\ 0 & 0.7 & 1.0 & \bigm| & 1.7 \\ 0 & 0 & -0.43 & \bigm| & -0.43 \end{bmatrix}$$

$$m_{32} = -0.43$$

Hence $x_1 = x_2 = x_3 = 1.00$.

The general algorithm for Gauss Elimination with interchanges is as follows. The obvious notation $a \leftrightarrow b$ means that the variables $a$ and $b$ have their values interchanged.

*Algorithm 5.4 Gauss elimination with interchanges, short version*

Elimination process:
For $k = 1, 2, \ldots, n - 1$:
   (i) Find pivot $a_{lk}$ such that $|a_{lk}| \geqslant |a_{ik}|$ for $i = k, \ldots, n (l \geqslant k)$
       Store row number $l(k)$ of pivot (for each $k$).
  (ii) $a_{kj} \leftrightarrow a_{lj}$ (for $l = l(k)$) for $j = k, \ldots, n$
       For $i = k + 1, \ldots, n$:
 (iii) $m_{ik} = -a_{ik}/a_{kk}$
  (iv) $a_{ij} := a_{ij} + m_{ik}a_{kj}$ (for $j = k + 1, \ldots, n$)
   (v) $b_k \leftrightarrow b_l$ (for $l = l(k)$)
  (vi) $b_i := b_i + m_{ik}b_k$
Back substitution:
 (vii) $x_n = b_n/a_{nn}$
       For $i = n - 1, \ldots, 1$:
$$x_i = \left( b_i - \sum_{j=i+1}^{n} a_{ij}x_j \right) \Big/ a_{ii}$$

Before proceeding to a BASIC program, we describe a further feature which may easily be added to Algorithm 5.4 to increase its flexibility.

### 5.4.1 New right-hand sides

One common requirement in practice is to be able to solve $\mathbf{Ax} = \mathbf{b}$ for more than one choice of $\mathbf{b}$, but with the same left-hand side $\mathbf{A}$. For example in structural engineering $\mathbf{A}$ might be a stiffness matrix, which is fixed for a given framework, and $\mathbf{b}$ might be an applied loading. Solutions $\mathbf{x}$ for the stresses might be required for a variety of different loadings $\mathbf{b}$. Another example might be the determination of the currents in the components of a fixed electrical circuit for different choices of applied voltages.

If $\mathbf{Ax} = \mathbf{b}$ has been solved once for a particular choice of $\mathbf{b}$, there is no need to perform all the calculations of Algorithm 5.4 to solve the equations for a second choice of $\mathbf{b}$. The multipliers $m_{ik}$, and the row interchanges $l(k)$ [interchange row $l$ and row $k$ in cycle $k$] have then already been determined, and the final upper triangular version $\mathbf{A}^{(n)}$ of $\mathbf{A}$ is already available for back-substitution. Thus we need only perform steps (v), (vi), and (vii) in Algorithm 5.4 with the newly introduced values of $b_1, \ldots, b_n$. The multiplications count corresponds to that for steps (iii) and (iv) of Algorithm 5.2 and is therefore just $n^2$ (approximately).

### 5.4.2 *Economy of storage*

In Gauss elimination we calculate multipliers $m_{ij}$ (for $i > j$), and, after the elimination process based on these multipliers, the final entries in the matrix are $a_{ij}^{(n)}$ (for $i \leqslant j$). Clearly the set of multipliers forms a matrix $\mathbf{M}$ with entries only below the diagonal:

$$\mathbf{M} = \begin{bmatrix} 0 & 0 & 0 & \dots & 0 & 0 \\ m_{21} & 0 & 0 & \dots & 0 & 0 \\ m_{31} & m_{32} & 0 & \dots & 0 & 0 \\ \vdots & & & & & \\ m_{n1} & m_{n2} & m_{n3} & \dots & m_{n,n-1} & 0 \end{bmatrix} \quad (5.21)$$

and the set of final entries $a_{ij}^{(n)}$ forms a matrix $\mathbf{U} = \mathbf{A}^{(n)}$ (the final version of $\mathbf{A}$ with entries only on or above the diagonal:

$$\mathbf{U} = \begin{bmatrix} a_{11}^{(n)} & a_{12}^{(n)} & \dots & a_{1n}^{(n)} \\ 0 & a_{22}^{(n)} & \dots & a_{2n}^{(n)} \\ 0 & 0 & & \\ \vdots & & & \\ 0 & 0 & \dots & a_{nn}^{(n)} \end{bmatrix} \quad (5.22)$$

There is therefore no need to allocate separate matrix storage locations for both $\mathbf{M}$ and $\mathbf{U}$, since together they can neatly fill a square array. Whenever an element $a_{ij}$ in $\mathbf{A}$ is eliminated, by use of the multiplier $m_{ij}$, we may store $m_{ij}$ in the space vacated in $\mathbf{A}$. Then at the end of the elimination process the matrix locations $\mathbf{A}$ will be occupied by $\mathbf{M}$ below the diagonal and $\mathbf{U}$ on and above the diagonal.

Program 5.4A is similar to Program 5.3 in structure, but incorporates the three improvements discussed above: row interchanges, any number of right-hand sides, and economy of storage. These are summarised in Algorithm 5.4A. The program is called GPLUS (short for Gauss Plus) and is a versatile Gauss elimination program.

*Algorithm 5.4A Gauss elimination with interchanges, versatile version*

This is identical to Algorithm 5.4, apart from the following addenda:
(a) $m_{ik}$ is stored in $a_{ik}$.
(b) control is returned to part (v) at the end of the program, so that a new $\mathbf{b}$ may be input and $\mathbf{A}\mathbf{x} = \mathbf{b}$ solved without repeating (i) to (iv).
(c) $|\mathbf{A}|$ and $\mathbf{r} = \mathbf{b} - \mathbf{A}\mathbf{x}$ are calculated as in Algorithm 5.3.

**Program 5.4A** GPLUS: Gauss elimination with interchanges, versatile version

```
LIST
GPLUS

10   REM-  GPLUS: SOLVES AX=B BY GAUSS ELIMINATION WITH INTERCHANGES
20   REM- ALSO CALCULATES RESIDUAL VECTOR B-AX AND DETERMINANT OF A
30   REM- THEN SOLVES FOR NEW R.H.S. B WITHOUT REPEATING CALCULATIONS ON A
40   REM- STORES MULTIPLIERS IN LOWER TRIANGLE OF A
50   DIM A(20,20),A1(20,20),X(20),B(20),B1(20),R(20)
60   PRINT "NUMBER OF EQUATIONS ";
70   INPUT N
80   PRINT "INPUT MATRIX A ROW BY ROW "
90   MAT INPUT A(N,N)
100  PRINT "INPUT VECTOR B"
110  MAT INPUT B(N)
120  REM- KEEPS COPIES OF ORIGINAL A,B IN A1,B1
130  MAT A1=A
140  MAT B1=B
150  REM- PERFORMS N-1 CYCLES OF ELIMINATIONS
160  REM- ACCUMULATES DETERMINANT OF A IN D1
170  D1=1
180  REM- CURRENT CYCLE NUMBER IS K
190  FOR K=1 TO N-1
200  REM- FINDS LARGEST ELEMENT IN COLUMN K
210  REM- RECORDS ROW NUMBER L(K) OF THIS ELEMENT
220  D=0
230  FOR I=K TO N
240  C=ABS(A(I,K))
250  IF D>=C THEN 280
260  D=C
270  L(K)=I
280  NEXT I
290  L1=L(K)
300  IF L1=K THEN 400
310  REM- CHANGES SIGN OF DETERMINANT
320  D1=-D1
330  REM- INTERCHANGES ROWS K AND L(K) OF A
340  FOR J=K TO N
350  D=A(K,J)
360  A(K,J)=A(L1,J)
370  A(L1,J)=D
380  NEXT J
390  REM- CALCULATES MULTIPLIER M(I,K) TO ELIMINATE A(I,K)
400  FOR I=K+1 TO N
410  M=-A(I,K)/A(K,K)
420  REM- STORES M IN NEWLY VACATED LOCATION A(I,K)
430  A(I,K)=M
440  REM- PERFORMS ROW OPERATIONS ON A
450  FOR J=K+1 TO N
460  A(I,J)=A(I,J)+M*A(K,J)
470  NEXT J
480  NEXT I
490  NEXT K
500  REM- CALCULATES DETERMINANT D1 OF A
510  FOR I=1 TO N
520  D1=D1*A(I,I)
530  NEXT I
540  PRINT "DETERMINANT OF A :";
550  PRINT D1
560  REM- INTERCHANGES ROWS K AND L(K) OF B
570  FOR K=1 TO N-1
580  L1=L(K)
590  IF L1=K THEN 640
600  D=B(K)
610  B(K)=B(L1)
620  B(L1)=D
630  REM- PERFORMS ROW OPERATIONS ON B
640  FOR I=K+1 TO N
650  M=A(I,K)
660  B(I)=B(I)+M*B(K)
670  NEXT I
680  NEXT K
690  REM- CALCULATES X(N),...,X(1) BY BACK SUBSTITUTION
700  X(N)=B(N)/A(N,N)
710  FOR I=N-1 TO 1 STEP -1
720  D=B(I)
```

```
730 FOR J=I+1 TO N
740 D=D-A(I,J)*X(J)
750 NEXT J
760 X(I)=D/A(I,I)
770 NEXT I
780 PRINT "SOLUTION X :"
790 MAT PRINT X(N)
800 REM- CALCULATES RESIDUAL VECTOR R=B-AX
810 FOR I=1 TO N
820 W=B1(I)
830 FOR J=1 TO N
840 W=W-A1(I,J)*X(J)
850 NEXT J
860 R(I)=W
870 NEXT I
880 PRINT "RESIDUAL VECTOR B-AX :"
890 MAT PRINT R(N)
900 REM-  SOLVES FOR A NEW B IF REQUIRED
910 PRINT "INPUT 1 FOR NEW B,INPUT 0 TO STOP"
920 INPUT C
930 IF C=0 THEN 980
940 PRINT "INPUT VECTOR B:"
950 MAT INPUT B(N)
960 MAT B1=B
970 GO TO 570
980 END
```

Ready

## Sample run 1

```
RUN
GPLUS

NUMBER OF EQUATIONS ? 2
INPUT MATRIX A ROW BY ROW
? .000001,1,1,1
INPUT VECTOR B
? 1.000001,2
DETERMINANT OF A :-.999999
SOLUTION X :
 1
 1
RESIDUAL VECTOR B-AX :
 0
-.596046E-07
INPUT 1 FOR NEW B,INPUT 0 TO STOP
? 0
Ready
```

## Sample run 2

```
RUN
GPLUS

NUMBER OF EQUATIONS ? 3
INPUT MATRIX A ROW BY ROW
? 1,1,1,1,2,5,1,5,10
INPUT VECTOR B
? 1,0,0
DETERMINANT OF A :-7
SOLUTION X :
 .714286
 .714286
-.428571
RESIDUAL VECTOR B-AX :
-.298023E-07
-.238419E-06
-.476837E-06
INPUT 1 FOR NEW B,INPUT 0 TO STOP
? 1
INPUT VECTOR B:
? 0,1,0
SOLUTION X :
 .714286
-1.28571
 .571429
```

```
RESIDUAL VECTOR B-AX :
  0
  0
 .476837E-06
INPUT 1 FOR NEW B,INPUT 0 TO STOP
? 1
INPUT VECTOR B:
? 0,0,1
SOLUTION X :
-.428571
 .571429
-.142857
RESIDUAL VECTOR B-AX :
 .149012E-07
  0
  0
INPUT 1 FOR NEW B,INPUT 0 TO STOP
?
  0
Ready
```

## Program notes

(1) There are many REM statements which should make the program self-explanatory. The program is based on Program 5.3, though the statements have been renumbered in sequence. The main blocks of new code are instructions 200–300 (to determine the pivot and its position), 330–380 and 560–620 (to perform row interchanges), 420–430 and 650 (to store $m_{ij}$ in $a_{ij}$ and later retrieve it), and 900–970 (to read in a new **b** and transfer control to instruction 570, thus avoiding repetition of calculations involving **A**).

(2) The main new BASIC codes that have been added to Program 5.3 are those for determining a pivot (the largest element in magnitude in a given column) and for interchanging elements in rows of **A** and **b**. However, simple codes were given in Section 3.3 of Reference 1 for performing such tasks, and the codes in instructions 200–300, 330–380 and 560–620 are based on these.

(3) The row $l$ containing the pivot (in cycle $k$) depends on $k$ and so it has been stored in the dimensioned variable L(K). This allows it to be retrieved in instructions 290 and more particularly 580, when it is needed in the interchanges. In order to dispense with the subscript (K) during loops such as 340–380, the variable L1 is set equal to L(K) and is used in its place. Note that L cannot be used as both a subscripted and unsubscripted variable, and therefore an instruction such as L = L(K) would not have worked.

(4) Note that row interchanges in **A** and **b** have been separated completely in the program. This is essential in order to introduce a new right-hand side, and perform row interchanges on **b** only (at instruction 570).

(5) Small residual vectors are generally obtained from the program (see for example the Sample runs).

(6) New right-hand sides may be introduced successfully, as is demonstrated in Sample run 2.

### 5.4.3 *Matrix inversion, application of Program 5.4A*

One natural application of the improved program 5.4A is in the calculation of the inverse matrix $\mathbf{A}^{-1}$ of a given matrix $\mathbf{A}$, since this in fact involves solving $\mathbf{Ax} = \mathbf{b}$ for a number of different right-hand sides $\mathbf{b}$.

Suppose the $n$ columns (in order) of the identity matrix $\mathbf{I}$ are denoted by

$$\mathbf{e}^{(1)}, \mathbf{e}^{(2)}, \ldots, \mathbf{e}^{(n)}$$

so that

$$\mathbf{e}^{(1)} = \begin{pmatrix} 1 \\ 0 \\ 0 \\ \vdots \\ 0 \end{pmatrix}, \ \mathbf{e}^{(2)} = \begin{pmatrix} 0 \\ 1 \\ 0 \\ \vdots \\ 0 \end{pmatrix}, \ \ldots, \ \mathbf{e}^{(n)} = \begin{pmatrix} 0 \\ 0 \\ 0 \\ \vdots \\ 1 \end{pmatrix} \tag{5.23}$$

and suppose that $\mathbf{x}^{(1)}, \mathbf{x}^{(2)}, \ldots, \mathbf{x}^{(n)}$ are the solutions of

$$\mathbf{Ax}^{(i)} = \mathbf{e}^{(i)} \ (i = 1, 2, \ldots, n) \tag{5.24}$$

Then the matrix $\mathbf{X}$, whose columns are $\mathbf{x}^{(1)}, \ldots, \mathbf{x}^{(n)}$, namely

$$\mathbf{X} = (\mathbf{x}^{(1)} | \mathbf{x}^{(2)} | \mathbf{x}^{(3)} | \ldots | \mathbf{x}^{(n)})$$

has the property that

$$\mathbf{AX} = (\mathbf{Ax}^{(1)} | \mathbf{Ax}^{(2)} | \ldots | \mathbf{Ax}^{(n)}) = (\mathbf{e}^{(1)} | \mathbf{e}^{(2)} | \ldots | \mathbf{e}^{(n)}) = \mathbf{I}$$

It follows that $\mathbf{X} = \mathbf{A}^{-1}$.

Thus the determination of $\mathbf{A}^{-1}$ reduces to the solution of (5.24) for the $n$ different right-hand sides $\mathbf{e}^{(i)}$ $(i = 1, \ldots, n)$, namely the columns (5.23) of $\mathbf{I}$.

For example, using exact arithmetic as an illustration, for

$$\mathbf{A} = \begin{pmatrix} 1 & 1 & 1 \\ 1 & 2 & 5 \\ 1 & 5 & 10 \end{pmatrix} \tag{5.25}$$

we solve $\mathbf{Ax} = \mathbf{b}$ for $\mathbf{b} = \begin{pmatrix} 1 \\ 0 \\ 0 \end{pmatrix}, \begin{pmatrix} 0 \\ 1 \\ 0 \end{pmatrix}, \begin{pmatrix} 0 \\ 0 \\ 1 \end{pmatrix}$ , respectively.

The Gauss elimination process is as follows, and for convenience all three right-hand sides are treated together and row interchanges are omitted.

$$\left( \begin{array}{ccc|ccc} 1 & 1 & 1 & 1 & 0 & 0 \\ 1 & 2 & 5 & 0 & 1 & 0 \\ 1 & 5 & 10 & 0 & 0 & 1 \end{array} \right) \rightarrow \left( \begin{array}{ccc|ccc} 1 & 1 & 1 & 1 & 0 & 0 \\ 0 & 1 & 4 & -1 & 1 & 0 \\ 0 & 4 & 9 & -1 & 0 & 1 \end{array} \right)$$

$$\rightarrow \begin{bmatrix} 1 & 1 & 1 & 1 & 0 & 0 \\ 0 & 1 & 4 & -1 & 1 & 0 \\ 0 & 0 & -7 & 3 & -4 & 1 \end{bmatrix}$$

Back-substituting for each of the three right-hand sides gives:

$$\mathbf{x}^{(1)} = \begin{bmatrix} 5/7 \\ 5/7 \\ -3/7 \end{bmatrix}, \mathbf{x}^{(2)} = \begin{bmatrix} 5/7 \\ -9/7 \\ 4/7 \end{bmatrix}, \mathbf{x}^{(3)} = \begin{bmatrix} -3/7 \\ 4/7 \\ -1/7 \end{bmatrix}$$

Hence

$$\mathbf{A}^{-1} = \tfrac{1}{7} \begin{bmatrix} 5 & 5 & -3 \\ 5 & -9 & 4 \\ -3 & 4 & -1 \end{bmatrix} \tag{5.26}$$

The *multiplications count* for determining $\mathbf{A}^{-1}$ consists of:
(a) Gauss elimination with one right-hand side: $\tfrac{1}{3}n^3$ multiplications.
(b) Gauss elimination with $n-1$ new right-hand sides ($n^2$ multiplications each): $(n-1)n^2$ multiplications.
The total is about $\tfrac{4}{3}n^3$ multiplications. We thus see that four times as much work is required to compute $\mathbf{A}^{-1}$ as to solve $\mathbf{Ax} = \mathbf{b}$, and this confirms the inefficiency of Algorithm 5.1.

The algorithm for $\mathbf{A}^{-1}$ is expressed formally as follows:

*Algorithm 5.4B The inverse matrix of* $\mathbf{A}$

(1) Solve $\mathbf{Ax} = \mathbf{b}$ by Algorithm 5.4A (Program 5.4A) with the $n$ right-hand sides

$$\mathbf{b} = \begin{bmatrix} 1 \\ 0 \\ 0 \\ \vdots \\ 0 \\ 0 \end{bmatrix}, \begin{bmatrix} 0 \\ 1 \\ 0 \\ \vdots \\ 0 \\ 0 \end{bmatrix}, \dots, \begin{bmatrix} 0 \\ 0 \\ 0 \\ \vdots \\ 0 \\ 1 \end{bmatrix}$$

(2) Place the respective solutions $\mathbf{x}^{(1)}, \mathbf{x}^{(2)}, \dots, \mathbf{x}^{(n)}$ side by side to form the matrix $\mathbf{A}^{-1}$:

$$\mathbf{A}^{-1} = (\mathbf{x}^{(1)}|\mathbf{x}^{(2)}|\dots|\mathbf{x}^{(n)})$$

This algorithm is tested on the matrix (5.25) in Sample run 2 of Program 5.4A.

The three right-hand sides are $(1\ 0\ 0)^{\mathrm{T}}$, $(0\ 1\ 0)^{\mathrm{T}}$, and $(0\ 0\ 1)^{\mathrm{T}}$ and the three solutions placed side by side form the matrix $\mathbf{A}^{-1}$, namely

$$\begin{bmatrix} 0.714286 & 0.714286 & -0.428571 \\ 0.714286 & -1.28571 & 0.571429 \\ -0.428571 & 0.571429 & -0.142857 \end{bmatrix}$$

This can be seen to be the correct inverse matrix (5.26).

### 5.4.4 *Other modifications*

A number of other types of modification of Gauss elimination have been proposed, which are concerned for example with more compact schemes and special types of matrices. For an elementary but detailed treatment of this area, the reader is referred to the excellent text of L. Fox (Reference 2). The discussion here will be limited to one special but important case, namely that in which the matrix **A** is symmetric.

## 5.5 Symmetric matrices

If the matrix **A** is symmetric, then the Gauss elimination method can be executed in half as many operations. Moreover only about half the matrix, namely the upper triangle, needs to be stored throughout the computation. However, in order to achieve these savings it is necessary either to make no row interchanges or to modify the row interchange scheme by choosing pivots on the diagonal of the matrix only.

Suppose first that we wish to simplify the short Gauss elimination method without interchanges, namely Algorithm 5.2.

Let us assume that at the start of cycle $(k)$, the square matrix in the right-hand corner of **A** (see (5.12)) is symmetric, i.e.

$$a_{ij}^{(k)} = a_{ji}^{(k)} (i = k, \ldots, n; i = k, \ldots, n) \qquad (5.27)$$

Then, from (5.13), the multipliers $m_{ik}$ are given by:

$$m_{ik} = -a_{ik}^{(k)}/a_{kk}^{(k)} \ (i = k + 1, \ldots, n)$$

and since $a_{ik}^{(k)} = a_{ki}^{(k)}$, it follows that

$$m_{ik} = -a_{ki}^{(k)}/a_{kk}^{(k)} \ (i = k + 1, \ldots, n) \qquad (5.28)$$

Note that (5.28) only involves entries in the upper triangle of **A**. The row operations on **A** in cycle $(k)$ of the elimination process take the form, from (5.14):

$$a_{ij}^{(k+1)} = a_{ij}^{(k)} + m_{ik}a_{kj}^{(k)} \ (i = k + 1, \ldots, n; j = k + 1, \ldots, n) \qquad (5.29)$$

Now

$$a_{ij}^{(k+1)} = a_{ij}^{(k)} - a_{ik}^{(k)}a_{kj}^{(k)}/a_{kk}^{(k)} \quad \text{by (5.28) and (5.29)}$$
$$= a_{ji}^{(k)} - a_{jk}^{(k)}a_{ki}^{(k)}/a_{kk}^{(k)} \quad \text{by (5.27)}$$
$$= a_{ji}^{(k+1)}$$

Hence (5.27) follows with $k$ replaced by $k + 1$. Since (5.27) holds for $k = 0$ by the symmetry of **A**, it holds for all $k$ by induction.

The consequence of this established symmetry is that we need only determine elements $a_{ij}^{(k)}$ for $j \geqslant i$. So (5.29) may be replaced by

$$a_{ij}^{(k+1)} = a_{ij}^{(k)} + m_{ik}a_{kj}^{(k)} \quad (i = k + 1, \ldots, n; j = i, \ldots, n) \qquad (5.30)$$

and this results in a saving of about half in all the arithmetic operations.

A simple (3 by 1) example will serve to illustrate the savings.

$$(\mathbf{A}|\mathbf{b}) = \begin{bmatrix} 1 & 2 & 5 \\ 2 & 2 & 4 \\ 5 & 4 & 5 \end{bmatrix} \begin{bmatrix} 8 \\ 8 \\ 14 \end{bmatrix} \rightarrow \begin{bmatrix} 1 & 2 & 5 \\ & 2 & 4 \\ & & 5 \end{bmatrix} \begin{bmatrix} 8 \\ 8 \\ 14 \end{bmatrix}$$

(Symmetric)    $m_{21} = -\frac{2}{1}, m_{31} = -\frac{5}{1}$

$$\rightarrow \begin{bmatrix} 1 & 2 & 5 \\ & -2 & -6 \\ & & -20 \end{bmatrix} \begin{bmatrix} 8 \\ -8 \\ -26 \end{bmatrix} \rightarrow \begin{bmatrix} 1 & 2 & 5 \\ & -2 & -6 \\ & & -2 \end{bmatrix} \begin{bmatrix} 8 \\ -8 \\ -2 \end{bmatrix}$$

$m_{32} = -(-6)/(-2)$

Then, by back substitution,

$$x_3 = 1, x_2 = 1, x_1 = 1$$

The symmetric algorithm and program, based on (5.28) and (5.30), are now given.

*Algorithm 5.5 Gauss elimination without interchanges, **A** symmetric*

For $k = 1, 2, \ldots, n - 1$:
For $i = k + 1, \ldots, n$:
(i) $m_{ik} = -a_{ki}/a_{kk}$
(ii) $a_{ij} := a_{ij} + m_{ik}a_{kj} (j = i, \ldots, n)$
(iii), (iv) as in Algorithm 5.2.

## Program 5.5 GYSMM: Gauss elimination without interchanges, symmetric matrix

```
LIST
GSYMM

10   REM- GSYMM: SOLVES AX=B FOR SYMMETRIC A
20   REM- USES GAUSS ELIMINATION WITHOUT PIVOTING
30   DIM A(20,20),X(20),B(20),R(20)
40   PRINT "NUMBER OF EQUATIONS";
50   INPUT N
60   PRINT "INPUT SYMMETRIC MATRIX A IN ROWS"
70   MAT INPUT A(N,N)
80   PRINT "INPUT VECTOR B"
90   MAT INPUT B(N)
100  REM- PERFORMS N-1 CYCLES OF ELIMINATIONS
110  FOR K=1 TO N-1
120  REM- CALCULATES MULTIPLIER M TO ELIMINATE A(I,K)
130  FOR I=K+1 TO N
140  M=-A(K,I)/A(K,K)
150  REM- PERFORMS ROW OPERATIONS ON A
160  FOR J=I TO N
170  A(I,J)=A(I,J)+M*A(K,J)
180  NEXT J
190  REM- PERFORMS ROW OPERATIONS ON B
200  B(I)=B(I)+M*B(K)
210  NEXT I
220  NEXT K
230  REM- CALCULATES X(N),...,X(1) BY BACK SUBSTN
240  X(N)=B(N)/A(N,N)
250  FOR I=N-1 TO 1 STEP -1
260  D=B(I)
270  FOR J=I+1 TO N
280  D=D-X(J)*A(I,J)
290  NEXT J
300  X(I)=D/A(I,I)
310  NEXT I
320  PRINT "SOLUTION X:"
330  MAT PRINT X(N)
340  END

Ready
```

*Sample run*

```
RUN
GSYMM

NUMBER OF EQUATIONS? 3
INPUT SYMMETRIC MATRIX A IN ROWS
? 1,1,1,1,2,5,1,5,10
INPUT VECTOR B
? 1,0,0
SOLUTION X:
 .714286
 .714286
-.428571
Ready
```

*Program notes*

(1) This program is essentially identical to Program 5.2 above apart from two changes. Firstly A(I, K) is replaced by A(K, I) in instruction 210, and secondly J = K + 1 is replaced by J = I in instruction 230. These two instructions have become instructions 140 and 160 in the new program, as a consequence of renumbering. One or two cosmetic changes have also been made.

(2) All Program Notes of Program 5.2 hold true for this program, with suitable amendments to Note 2 to account for the renumbering of instructions.

### 5.5.1 *Row and column interchanges for symmetric matrices*

If we wish to make row interchanges, so as to ensure that no pivots are zero or close to zero, then we must use 'diagonal pivoting' in order to preserve the symmetry of $\mathbf{A}$. This means that, in cycle $(k)$, the largest element in magnitude amongst $a_{kk}^{(k)}, a_{k+1,k+1}^{(k)}, \ldots, a_{nn}^{(k)}$ must be found and chosen as pivot. If $a_{ll}^{(k)}$ is the pivot, then rows $k$ and $l$ of $\mathbf{A}$ are interchanged and so are columns $k$ and $l$ of $\mathbf{A}$. It is easy to verify that this preserves the symmetry of $\mathbf{A}$ and brings the pivot into the position originally held by $a_{kk}^{(k)}$, so that row operations can then proceed as in Algorithm 5.5.

The consequence of interchanging columns $k$ and $l$, say, in $\mathbf{A}$ is that it becomes necessary also to interchange the components $x_k$ and $x_l$ of the solution $\mathbf{x}$. This is most easily achieved by keeping a record of the interchange $l(k)$ of the suffix $l$ with the suffix $k$ for each $k$, and then performing these interchanges of suffices in reverse order to $\{x_i\}$ after the Gauss elimination process has been completed.

We shall not give a BASIC program which incorporates diagonal pivoting, but rather will leave this as exercise to the reader (see Problem 10). The necessary algorithm may be summarised as follows.

*Algorithm 5.5A Gauss elimination with diagonal pivoting, $\mathbf{A}$ symmetric*

As Algorithm 5.5, with the following additions:
For cycles $(k)$ $(k = 1, \ldots, n-1)$:
(a) Determine $a_{ll}$ $(k \leqslant l \leqslant n)$ such that $|a_{ll}| \geqslant |a_{ii}|$ $(i = k, \ldots, n)$
(b) Interchanges: $a_{kj} \leftrightarrow a_{lj}$ $(j = k, \ldots, n-1)$
$$b_k \leftrightarrow b_l$$
$$a_{ik} \leftrightarrow a_{il} \ (i = k, \ldots, n)$$
$$x_i \leftrightarrow x_l$$

### 5.5.2 *A practical application: least squares*

Program 5.5 (without pivoting) can be used in one important practical application. Suppose that it is required to solve the *overdetermined* system of equations

$$\mathbf{Bx} = \mathbf{y} \tag{5.31}$$

where $\mathbf{B}$ is $m$ by $n$, $\mathbf{x}$ is $n$ by 1 and $\mathbf{y}$ is $m$ by 1 $(m \geqslant n)$. Here the number of equations exceeds the number of unknowns, and so there are in general no exact solutions. However, an approximate solution may be obtained by the method of least squares. In this method, we minimise with respect to the unknowns $x_1$, $x_2$, ..., $x_n$ the sum of squares of the errors in each individual equation, namely the quantity

$$\sum_{i=1}^{m} [(b_{i1}x_1 + \ldots + b_{in}x_n) - y_i]^2 \qquad (5.32)$$

The mathematical details will be left as an exercise (Problem 11), but the method has the effect of premultiplying both sides of (5.31) by $\mathbf{B}^T$ to give

$$\mathbf{Ax} = \mathbf{b} \qquad (5.33)$$

where

$$\mathbf{A} = \mathbf{B}^T\mathbf{B} \text{ and } \mathbf{b} = \mathbf{B}^T\mathbf{y}$$

Clearly the resulting well-determined $n$ by $n$ system (5.33) has on its left-hand side a symmetric matrix (often called the 'normal matrix' of (5.31)), namely

$$\mathbf{A} = \mathbf{B}^T\mathbf{B}.$$

This matrix is also 'positive definite' (i.e. for any $n$ by 1 vector $\mathbf{z}$, $\mathbf{z}^T\mathbf{Az} \geqslant 0$ with equality if and only if $\mathbf{z} = 0$). More importantly it is easy to see that no diagonal elements of $\mathbf{A}$ can be zero, and hence Program 5.5 is applicable.

## 5.6 Summary of Gauss elimination algorithms

In this chapter we have discussed a logical progression of no less than three categories of programs for Gauss elimination, and so it would be a good idea to summarise this progression here.
*Category 1:* Program 5.2 is a short program which computes a 'solution' $\mathbf{x}$. However, in some cases the computed solution neither satisfies $\mathbf{Ax} = \mathbf{b}$ nor is close to the true solution. The program does not tell us this!
*Category 2:* Program 5.3 also computes $|\mathbf{A}|$ and $\mathbf{r}$ (residual). If $\mathbf{r}$ is small then we know that the equations $\mathbf{Ax} = \mathbf{b}$ have been satisfied. If $|\mathbf{A}|$ is small, then $\mathbf{x}$ may not be close to the true solution. (In general $\mathbf{x}$ is only close to the true solution if $\mathbf{Ax} = \mathbf{b}$ is well-conditioned.)
*Category 3:* Program 5.4A will normally produce a solution $\mathbf{x}$ for which the residual $\mathbf{r}$ is 'zero' to computer (single precision) accuracy, so that the equations $\mathbf{Ax} = \mathbf{b}$ are effectively satisfied. However, as in Program 5.3, $\mathbf{x}$ may still not be close to the true solution, especially if $|\mathbf{A}|$ is small.

It should therefore be clear to you, the reader, that the solution of a system of simultaneous linear equations is a problem which should not be tackled blindly! If you decide to use a library program or someone else's program you would be well advised to ascertain, if you can, to which of the above three categories the program belongs. The chances are that it will belong to Category 3, though there is always the danger that it may belong to Category 2 or, worst still, Category 1. In any case, the remarks that we have made about the nature of the computed solution should be thoroughly digested and understood.

## 5.7 Further reading

The 'Category 3' Algorithm 5.4A, which may not always produce an acceptable solution $\mathbf{x}$, can be improved further. Specifically the method of 'iterative refinement' (see Reference 3), which uses 'double precision' arithmetic in a small part of the calculation can lead to residuals which are zero to 'double precision' accuracy. Unfortunately such a method cannot be implemented in a single BASIC program on most computer systems since it is not normally possible to perform one part of a BASIC program in single precision and another part in double precision.

## 5.8 References

1. Mason, J.C., *BASIC Numerical Mathematics*, Butterworths (1983).
2. Fox, L., *An Introduction to Numerical Linear Algebra*, Oxford University Press (1964).
3. Forsythe, G.E. and Moler, C.B., *Computer Solution of Linear Algebraic Systems*, Prentice-Hall, New Jersey (1967).

## PROBLEMS

**(5.1)** Test Program 5.1 on the following system of equations, and comment on the uniqueness or otherwise of the true solutions.

(i)
$$\begin{bmatrix} 1 & 2 & 3 \\ 4 & 5 & 6 \\ 7 & 8 & 9 \end{bmatrix} \begin{bmatrix} x_1 \\ x_2 \\ x_3 \end{bmatrix} = \begin{bmatrix} 6 \\ 15 \\ 24 \end{bmatrix}$$

(ii)
$$\begin{bmatrix} 1 & 1 & 1 \\ 1 & 2 & 3 \\ 4 & 5 & 6 \end{bmatrix} \begin{bmatrix} x_1 \\ x_2 \\ x_3 \end{bmatrix} = \begin{bmatrix} 3 \\ 6 \\ 9 \end{bmatrix}$$

(iii) $\begin{bmatrix} 0 & 1 & 1 \\ 1 & 0 & 1 \\ 1 & 1 & 0 \end{bmatrix} \begin{bmatrix} x_1 \\ x_2 \\ x_3 \end{bmatrix} = \begin{bmatrix} 1 \\ 2 \\ 3 \end{bmatrix}$

(iv) $\begin{bmatrix} 1 & 1/2 & 1/3 \\ 1/2 & 1/3 & 1/4 \\ 1/3 & 1/4 & 1/5 \end{bmatrix} \begin{bmatrix} x_1 \\ x_2 \\ x_3 \end{bmatrix} = \begin{bmatrix} 11/6 \\ 13/12 \\ 47/60 \end{bmatrix}$

Test the conditioning of systems (iii) and (iv) by making changes (e.g. 0.01) to some of the coefficients and observing the consequent changes to the solution. You should find (iii) to be well-conditioned and (iv) to be ill-conditioned.

**(5.2)** Write a program, using the BASIC matrix routines, to calculate $A^2$, $A^3$ ..., and $(I - A)^{-1}$ for the matrix

$$A = \begin{bmatrix} 0.05 & 0.05 \\ -0.05 & 0.05 \end{bmatrix}.$$

Hence test the validity of the 'binomial expansion' for matrices:

$$(I - A)^{-1} = I + A + A^2 + A^3 + \cdots$$

which is known to hold for matrices $A$ with suitably small elements. Here $A^2 = A.A$, $A^3 = A^2.A$, etc.

**(5.3)** Test Program 5.2 (or 5.3) on your computer for the system

$$\begin{bmatrix} \varepsilon & 1 \\ 1 & 1 \end{bmatrix} \begin{bmatrix} x_1 \\ x_2 \end{bmatrix} = \begin{bmatrix} 1 + \varepsilon \\ 2 \end{bmatrix},$$

where $\varepsilon$ is a specified small number. (The true solution is $x_1 = x_2 = 1$.)

Determine the smallest value of $\varepsilon$ for which the program produces an answer (but below which the computer gives an error message). How large is the error in the computed solution for this $\varepsilon$?

**(5.4)** Define

$$A = \begin{bmatrix} 1 & 1/2 & 1/3 & \cdots & 1/n \\ 1/2 & 1/3 & 1/4 & \cdots & 1/(n+1) \\ 1/3 & 1/4 & 1/5 & \cdots & 1/(n+2) \\ \vdots & & & & \vdots \\ 1/n & 1/(n+1) & 1/(n+2) & \cdots & 1/(2n-1) \end{bmatrix}$$

and $b = A (1\ 1\ 1\ \cdots\ 1)^T$, (i.e. $b$ is the $n$ by 1 vector, each of whose elements is the sum of the entries in the corresponding row of $A$). Use Program 5.3 or 5.4A to solve for $x$ the system

$$Ax = b$$

for a number of increasing values of $n$, where the coefficients in $A$, $b$ are input (or calculated) as decimal numbers. What is the first value of $n$ at which ill-conditioning leads to a solution $\mathbf{x}$ with no correct figures? [The true $\mathbf{x}$ is $(1\ 1\ 1\ \ldots\ 1)^{\mathrm{T}}$.]

**(5.5)** By suitably modifying Program 5.4A, write a program to implement Algorithm 5.4B (to calculate the inverse matrix $\mathbf{A}^{-1}$ for a given $\mathbf{A}$).

Test the program on the matrix (5.25) for which the solution is (5.26).

**(5.6)** The following formula is known for $\mathbf{A}^{-1}$:

$$\mathbf{A}^{-1} = (\text{adj } \mathbf{A})/|\mathbf{A}|.$$

(See Section 3.7.) Why isn't this method used in standard computer programs? How many arithmetic operations are involved if $\mathbf{A}$ is $n$ by $n$?

**(5.7)** If $\mathbf{A}$ is a given $n$ by $n$ matrix and $\mathbf{B}$ is a given $n$ by $p$ matrix deduce an algorithm for determining the unknown $n$ by $p$ matrix $\mathbf{X}$ such that

$$\mathbf{AX} = \mathbf{B}$$

(This algorithm includes Algorithm 5.4B as a special case when $\mathbf{B} = \mathbf{I}$ and $n = p$.) Hint: Use a similar method to that of Section 5.4.3, but do *not* compute $\mathbf{A}^{-1}$.

Hence solve for $\mathbf{X}$ the matrix equation

$$\begin{bmatrix} 1 & 1 & 1 \\ 1 & 2 & 5 \\ 1 & 5 & 10 \end{bmatrix} . \mathbf{X} = \begin{bmatrix} 1 & 1 \\ 0 & 1 \\ 1 & 0 \end{bmatrix}$$

using one of the Programs 5.2, 5.3, 5.4A, 5.5.

**(5.8)** For the particular system (5.5) above (for which the Gauss elimination method was applied in detail in Section 5.3.2) verify that the left-hand side matrix $\mathbf{A}$ may be expressed as

$$\mathbf{A} = \mathbf{LU}$$

with $\mathbf{L} = \mathbf{I} - \mathbf{M}$, where $\mathbf{M}$ is the matrix of multipliers (see 5.21)), and $\mathbf{U} = \mathbf{A}^{(n)}$, the final version of $\mathbf{A}$ after the elimination procedure. Note: $\mathbf{L}$ has ones on its diagonal, and entries only on or below the diagonal. Such a matrix is called 'unit lower triangular'.

**(5.9)** Assuming that the relation

$$\mathbf{A} = \mathbf{L}.\mathbf{U}$$

always holds true for *any* matrix $\mathbf{A}$ (where $\mathbf{L}$ and $\mathbf{U}$ are defined as $\mathbf{I} - \mathbf{M}$ and $\mathbf{A}^{(n)}$, respectively, as in Problem 8), prove that the Gauss elimination method for solving $\mathbf{Ax} = \mathbf{b}$ can be expressed in the following way:

(i) Determine a unit lower triangular matrix **L** and an upper triangular matrix **U** such that

$$\mathbf{A} = \mathbf{L.U}$$

(ii) Solve the system

$$\mathbf{Lf} = \mathbf{b} \text{ for } \mathbf{f} \text{ (by forward substitution)}$$

(iii) Solve the system

$$\mathbf{Ux} = \mathbf{f} \text{ for } \mathbf{x} \text{ (by backward substitution)}$$

Use this new form of Gauss elimination to solve the system (5.5).
**(5.10)** Write a program to implement Algorithm 5.5A (symmetric Gauss elimination with diagonal pivoting). Test your program on the system

$$\begin{pmatrix} 0.00001 & 1 & 1 \\ 1 & 1 & 2 \\ 1 & 2 & 4 \end{pmatrix} \begin{pmatrix} x_1 \\ x_2 \\ x_3 \end{pmatrix} = \begin{pmatrix} 2.00001 \\ 4 \\ 7 \end{pmatrix}$$

Why does Algorithm 5.5 give inaccurate results for this problem?
**(5.11)**
(i) The 'least squares' solution **x** of the $m$ equations

$$b_{i1}x_1 + \ldots + b_{in}x_n = y_i \ (i = 1, \ldots, m) \ (m \geqslant n)$$

(i.e. **Bx** = **y**) is obtained by minimising the sum $S$ of the squares of the errors, namely

$$S = \sum_{i=1}^{m} [(b_{i1}x_1 + \ldots + b_{in}x_n) - y_i]^2$$

with respect to the variables $x_1, \ldots, x_n$.
Prove that the least squares solution **x** is given by the symmetric system

$$\mathbf{B^T B x} = \mathbf{B^T y}$$

Hint: The stationary points of $F(x_1, \ldots, x_n) = 0$ satisfy
$$\frac{\partial F}{\partial x_1} = \frac{\partial F}{\partial x_2} = \ldots = \frac{\partial F}{\partial x_n} = 0.$$
(ii) Hence determine (by hand) the least squares solution of the equations

$$\begin{pmatrix} 1 & 1 \\ 1 & 2 \\ 2 & 3 \end{pmatrix} \begin{pmatrix} x_1 \\ x_2 \end{pmatrix} = \begin{pmatrix} 2.1 \\ 3.1 \\ 6.1 \end{pmatrix}$$

What value does $\sqrt{S/m}$, the root mean square error, have in this case? [Answers: $\mathbf{x} = (1.4\ 1)^{\text{T}}$, $\sqrt{S/m} = 0.3$.]

**(5.12)** Write a program based on Programs 4.1 and 5.5 to determine the least squares solution $\mathbf{x}$ of an overdetermined system $\mathbf{Bx} = \mathbf{f}$, where $\mathbf{B}$ and $\mathbf{f}$ are given $m \times n$ and $m \times 1$ matrices, respectively, and to calculate the resulting root mean square error.

Test this program on the system in Problem 11 and also on the system

$$\begin{pmatrix} 1 & 1 & 1 \\ 1 & 2 & 3 \\ 2 & 3 & 6 \\ 2 & 4 & 7 \end{pmatrix} \begin{pmatrix} x_1 \\ x_2 \\ x_3 \end{pmatrix} = \begin{pmatrix} 3.1 \\ 6.1 \\ 11.1 \\ 13.1 \end{pmatrix}.$$

# Chapter 6

# Matrix calculations: iteration methods

## ESSENTIAL THEORY

### 6.1 Introduction

In the last chapter the linear algebraic system

$$\mathbf{Ax} = \mathbf{b} \tag{6.1}$$

was solved by the Gauss elimination method, a *direct* method of determining $\mathbf{x}$ in a finite number of steps. However, it is also possible to adopt methods in which successive approximations are obtained to $\mathbf{x}$ by an *iteration* (or iterative process). The aim is then that each approximation should be better than the previous one, so that the sequence of approximations *converges* to the true solution $\mathbf{x}$. Unlike a direct method, an iteration method is subject to *truncation error*, measured in terms of the difference between the approximate and true solutions. Iteration methods are also considered in Chapter 4 of Reference 1 in connection with a single nonlinear algebraic equation, and the reader might find it instructive to note the similarity of approach.

There are three main advantages of iteration methods over direct methods for (6.1) which we now indicate briefly. Firstly, there are a large number of matrices $\mathbf{A}$ arising in physical problems in which many elements are zero, and for such 'sparse matrices' iteration methods can be very efficient. In particular non-zero elements frequently occur in a regular pattern in the matrix, and it is advantageous to retain this structure during the solution. The Gauss elimination process as it progresses, tends to 'fill in' the upper triangular part of $\mathbf{A}$ with non-zero entries, even in places where zeros occurred initially, whereas the iteration methods of Jacobi and Gauss–Seidel do not fill in the matrix in this way.

Secondly, iteration methods are very simple to operate and program compared with the Gauss elimination method. Most iteration methods do not need to start from a good approximation to $\mathbf{x}$, and indeed often almost any $\mathbf{x}$ can be used initially. Each iteration needs at most about $n^2$ multiplications and so the method can be as efficient as Gauss elimination provided that not too many iterations are involved.

Thirdly, iteration methods are effectively not subject to rounding error, since solutions are 'corrected' at each iteration. Hence the accuracy of a computed solution may be confirmed by comparing it with the solution computed at the previous iteration.

## 6.2 Iteration method of Jacobi

Consider for example the system of four equations

$$\left.\begin{array}{rrrrr} -30x_1 + & x_2 + & x_3 + & x_4 = & -30 \\ x_1 - & 30x_2 + & x_3 + & x_4 = & 0 \\ x_1 + & x_2 - & 30x_3 + & x_4 = & -30 \\ x_1 + & x_2 + & x_3 - & 30x_4 = & 0 \end{array}\right\} \quad (6.2)$$

These equations can obviously be rewritten so as to isolate $x_1$, $x_2$, $x_3$, $x_4$, respectively, on the left-hand side as follows:

$$\left.\begin{array}{rr} -30x_1 = & -30 - x_2 - x_3 - x_4 \\ -30x_2 = & 0 - x_1 - x_3 - x_4 \\ -30x_3 = & -30 - x_1 - x_2 - x_4 \\ -30x_4 = & 0 - x_1 - x_2 - x_3 \end{array}\right\} \quad (6.3)$$

It is fairly clear, by inspection, that the solution is not too far from the values $x_1 = x_3 = 1, x_2 = x_4 = 0$. So we take these values as initial approximations and write

$$\mathbf{x}^{(0)} = \begin{bmatrix} x_1^{(0)} \\ x_2^{(0)} \\ x_3^{(0)} \\ x_4^{(0)} \end{bmatrix} = \begin{bmatrix} 1 \\ 0 \\ 1 \\ 0 \end{bmatrix} \quad (6.4)$$

the superscript (0) denoting the number of the iteration (i.e. iteration zero in this case).

The iteration now proceeds as follows. Insert the values (6.4) on the right-hand side of (6.3), and solve for new values $x_1^{(1)}, \ldots, x_4^{(1)}$:

$$\left.\begin{array}{rl} -30x_1^{(1)} = & -30 - 0 - 1 - 0 = -31 \\ -30x_2^{(1)} = & 0 - 1 - 1 - 0 = -2 \\ -30x_3^{(1)} = & -30 - 1 - 0 - 0 = -31 \\ -30x_4^{(1)} = & 0 - 1 - 0 - 1 = -2 \end{array}\right\} \Rightarrow$$

$$\begin{bmatrix} x_1 \\ x_2 \\ x_3 \\ x_4 \end{bmatrix}^{(1)} = \begin{bmatrix} 1.033 \\ .067 \\ 1.033 \\ .067 \end{bmatrix} \quad (6.5)$$

Repeat this calculation with $x_1^{(1)}, \ldots, x_4^{(1)}$ inserted on the right:

$$
\left.
\begin{array}{l}
-30x_1^{(2)} = -30 - .067 - 1.033 - .067 \\
-30x_2^{(2)} = \phantom{-3}0 - 1.033 - 1.033 - .067 \\
-30x_3^{(2)} = -30 - 1.033 - .067 - .067 \\
-30x_4^{(2)} = \phantom{-3}0 - 1.033 - .067 - 1.033
\end{array}
\right\} \Rightarrow
$$

$$
\begin{bmatrix} x_1 \\ x_2 \\ x_3 \\ x_4 \end{bmatrix}^{(2)} =
\begin{bmatrix} 1.0389 \\ .0711 \\ 1.0389 \\ .0711 \end{bmatrix} \tag{6.6}
$$

This process can be repeated as often as required; in this example we observe that one additional decimal place of accuracy is obtained at each iteration, and indeed (6.4), (6.5), (6.6) are, respectively, correct to one, two, and three decimal places.

In the general case the problem is to solve $n$ equations of the form

$$
\left.
\begin{array}{l}
a_{11}x_1 + a_{12}x_2 + \ldots + a_{1n}x_n = b_1 \\
a_{21}x_1 + a_{22}x_2 + \ldots + a_{2n}x_n = b_2 \\
\phantom{a_{11}x_1}\vdots \phantom{aaaaaaaaaaaaaa} \vdots \\
a_{n1}x \phantom{_1} + a_{n2}x_2 + \ldots + a_{nn}x_n = b_n
\end{array}
\right\} \tag{6.7}
$$

Using the same method as above, new values $x_1^{(1)}, \ldots, x_n^{(1)}$ are calculated from initial guesses $x_1^{(0)}, \ldots, x_n^{(0)}$ by isolating $x_1, \ldots, x_n$, respectively, on the left-hand side:

$$
\left.
\begin{array}{l}
a_{11}x_1^{(1)} = b_1 - a_{12}x_2^{(0)} - a_{13}x_3^{(0)} - \ldots - a_{1n}x_n^{(0)} \\
a_{22}x_2^{(1)} = b_2 - a_{21}x_1^{(0)} - a_{23}x_3^{(0)} - \ldots - a_{2n}x_n^{(0)} \\
\phantom{a_{11}x_1}\vdots \\
a_{nn}x_n^{(1)} = b_n - a_{n1}x_1^{(0)} - a_{n2}x_2^{(0)} - \ldots - a_{n,n-1}x_{n-1}^{(0)}
\end{array}
\right\} \tag{6.8}
$$

It is not difficult to deduce the general formula for determining $x_1, \ldots, x_n$ at iteration $(k)$ from $x_1, \ldots, x_n$ at iteration $(k-1)$ in the form (6.9) below. The complete algorithm is therefore as follows:

*Algorithm 6.1 Jacobi's method*

   (i) Choose initial values $x_1^{(0)}, \ldots, x_n^{(0)}$. If no choice is obvious, take $x_1^{(0)} = \ldots = x_n^{(0)} = 1$.

  (ii) For $k = 1, 2, \ldots$ calculate

$$
x_i^{(k)} = \left( b_i - \sum_{j=1}^{i-1} a_{ij}x_j^{(k-1)} - \sum_{j=i+1}^{n} a_{ij}x_j^{(k-1)} \right) \Big/ a_{ii} \quad (i = 1, \ldots, n) \tag{6.9}
$$

(Note: the first $\Sigma$ is omitted for $i = 1$, and the second $\Sigma$ is omitted for $i = n$.)

(iii) Terminate iteration on $k$ when $|x_i^{(k)} - x_i^{(k-1)}|$ is sufficiently small for every $i$.

The BASIC program, based on Algorithm 6.1, is given below. Note that no elements $a_{ij}$ in **A** have been assumed to be zero, and so no advantage has been taken of any special structure in **A**.

**Program 6.1** JACOBI: Jacobi iteration

```
LIST
JACOBI

10    REM- JACOBI: SOLVES AX=B BY JACOBI ITERATION
20    REM- Z DENOTES OLD SOLUTION , X DENOTES NEW SOLUTION.
30    DIM A(20,20)
40    DIM X(20),B(20),Z(20)
50    PRINT "NUMBER OF EQUATIONS ";
60    INPUT N
70    PRINT "INPUT A ROW BY ROW "
80    MAT INPUT A(N,N)
90    PRINT "INPUT B"
100   MAT INPUT B(N)
110   REM- INITIALIZE Z AS ALL ONES
120   FOR I=1 TO N
130   Z(I)=1
140   NEXT I
150   PRINT "ABSOLUTE ACCURACY REQUIRED ";
160   INPUT E
170   PRINT "MAXIMUM NUMBER OF ITERATIONS REQUIRED ";
180   INPUT M
190   PRINT "DO YOU WISH TO CHOOSE AN INITIAL VECTOR X ?"
200   PRINT "TYPE 0 FOR NO , TYPE 1 FOR YES"
210   INPUT C
220   IF C=0 THEN 250
230   PRINT "INPUT INITIAL VECTOR X"
240   MAT INPUT Z(N)
250   K=0
260   REM- ITERATION (K) STARTS HERE
270   REM- CALCULATE X(I) FROM Z(I)
280   REM- CURRENT ACCURACY =D = MAX ABS (X(I)-Z(I))
290   K=K+1
300   D=0
310   PRINT
320   PRINT "ITERATION",K
330   FOR I=1 TO N
340   C=B(I)
350   IF I=1 THEN 390
360   FOR J=1 TO I-1
370   C=C-A(I,J)*Z(J)
380   NEXT J
390   IF I=N THEN 430
400   FOR J=I+1 TO N
410   C=C-A(I,J)*Z(J)
420   NEXT J
430   X(I)=C/A(I,I)
440   F=ABS(X(I)-Z(I))
450   IF D>F THEN 470
460   D=F
470   NEXT I
480   PRINT "VECTOR X :"
490   FOR I=1 TO N
500   Z(I)=X(I)
510   PRINT X(I),
520   NEXT I
530   PRINT
540   REM- TEST ACCURACY AND NUMBER OF ITERATIONS
550   IF D<=E THEN 580
560   IF K>=M THEN 600
570   GO TO 290
580   PRINT "ABSOLUTE ACCURACY ATTAINED"
590   GO TO 610
600   PRINT "MAX NO OF ITERATIONS REACHED"
610   END

Ready
```

## Sample run 1

```
RUN
JACOBI

NUMBER OF EQUATIONS ? 4
INPUT A ROW BY ROW
? -30,1,1,1,1,-30,1,1,1,1,-30,1,1,1,1,-30
INPUT B
? -30,0,-30,0
ABSOLUTE ACCURACY REQUIRED ? .00001
MAXIMUM NUMBER OF ITERATIONS REQUIRED ? 10
DO YOU WISH TO CHOOSE AN INITIAL VECTOR X ?
TYPE 0 FOR NO , TYPE 1 FOR YES
? 1
INPUT INITIAL VECTOR X
? 1,0,1,0

ITERATION        1
VECTOR X :
  1.03333        .666667E-01    1.03333       .666667E-01

ITERATION        2
VECTOR X :
  1.03889        .711111E-01    1.03889       .711111E-01

ITERATION        3
VECTOR X :
  1.03937        .716296E-01    1.03937       .716296E-01

ITERATION        4
VECTOR X :
  1.03942        .071679        1.03942       .071679

ITERATION        5
VECTOR X :
  1.03943        .071684        1.03943       .071684
ABSOLUTE ACCURACY ATTAINED
Ready
```

## Sample run 2

```
RUN
JACOBI

NUMBER OF EQUATIONS ? 3
INPUT A ROW BY ROW
? 2,1,4,    1,2,4,    4,4,11
INPUT B
? 7,7,19
ABSOLUTE ACCURACY REQUIRED ? .01
MAXIMUM NUMBER OF ITERATIONS REQUIRED ? 7
DO YOU WISH TO CHOOSE AN INITIAL VECTOR X ?
TYPE 0 FOR NO , TYPE 1 FOR YES
? 1
INPUT INITIAL VECTOR X
? .9,.9,.9

ITERATION        1
VECTOR X :
  1.25           1.25          1.07273

ITERATION        2
VECTOR X :
  .729545        .729545       .818182

ITERATION        3
VECTOR X :
  1.49886        1.49886       1.19669

ITERATION        4
VECTOR X :
  .357179        .357179       .63719

ITERATION        5
VECTOR X :
  2.04703        2.04703       1.46751
```

```
ITERATION      6
VECTOR X :
-.458527       -.458527        .238523

ITERATION      7
VECTOR X :
 3.25222        3.25222        2.06075
MAX NO OF ITERATIONS REACHED
Ready
```

## Program notes

(1)  If matrix routines are not available, the MAT INPUT statements 80 and 100 may be replaced by loops of individual INPUT statements (see Program 5.2 notes).

(2)  For each cycle ($k$), Z(I) is the value of $x_i^{(k-1)}$ and X(I) is the value of $x_i^{(k)}$.

(3)  The algorithm halts if $\max_i |x_i^{(k)} - x_i^{(k-1)}| \leqslant \varepsilon$, i.e.

$$D = \max_I |X(I) - Z(I)| \leqslant E.$$ The required absolute accuracy $\varepsilon$ (i.e. E) is input in instruction 160, D is calculated in instructions 300, 330, 440–470, and the inequality $D \leqslant E$ is tested in instruction 550.

(4)  As a safety measure, the algorithm automatically halts after a specified number M of iterations has been reached (instruction 560).

(5)  If the user does not choose to input an initial vector **x** (after being questioned by instructions 190 and 200), then the program automatically takes $\mathbf{x}^{(0)} = (1 \ 1 \ 1 \ \dots \ 1)^T$, which is set in instructions 120–140.

(6)  The instruction 530 PRINT has the effect of moving the output onto a new line. This is necessary following the instruction 500 PRINT X(I), which outputs a value without changing lines. We may also of course use a new line (as in 310) to provide spacing in the output.

(7)  In Sample run 1 the program successfully solves example (6.2) above where **A** is strictly diagonally dominant.

(8)  In Sample run 2, the algorithm is clearly *not* converging. However here the chosen matrix **A** is *not* strictly diagonally dominant (see Section 6.2.2):

$$\mathbf{A} = \begin{bmatrix} 2 & 1 & 4 \\ 1 & 2 & 4 \\ 4 & 4 & 11 \end{bmatrix}$$

### 6.2.1  Special structure

Program 6.1 can be modified to take account of special structure in **A**. In particular this is easily achieved if **A** has the same number $p$, say, of adjacent non-zero entries to both right and left of the diagonal entry in each row, and then **A** is said to have band width

$2p + 1$. Such matrices frequently occur in connection with the solution of boundary value problems for differential equations. In particular, when $p = 1$, the matrix **A** is called tridiagonal, and an important example takes the following form:

$$\mathbf{A} = \begin{bmatrix} v & w & 0 & 0 & 0 & \ldots & 0 & 0 & 0 \\ u & v & w & 0 & 0 & \ldots & 0 & 0 & 0 \\ 0 & u & v & w & 0 & \ldots & 0 & 0 & 0 \\ \vdots & & & & & & & & \\ 0 & 0 & 0 & 0 & 0 & \ldots & u & v & w \\ 0 & 0 & 0 & 0 & 0 & \ldots & 0 & u & v \end{bmatrix} \qquad (6.10)$$

where $u$, $v$, $w$ are specified constants.

[This matrix occurs in particular in the solution by finite differences of second order ordinary differential equations with constant coefficients, and also in the solution of the heat conduction equation by the Crank–Nicolson method.]

For the specific system (6.10), it is only necessary to input and store $u$, $v$, and $w$ in order to specify **A** and define the Jacobi iteration. This system is solved in Program 6.2 which follows.

It should be pointed out that there are more efficient methods for solving tridiagonal systems, because of their very simple form. For example a fast direct algorithm, in which the number of arithmetic operations is proportional to the order $n$ of **A**, is described by G.D. Smith (Reference 2, pp. 20–22). However, our main aim here is to illustrate the general use of iteration methods for sparse matrices **A**.

**Program 6.2** JACTRI: Jacobi iteration for tridiagonal matrix

```
LIST
JACTRI

10   REM- JACTRI: SOLVES AX=B BY JACOBI ITERATION
15   REM- A IS TRIDIAGONAL WITH 3 CONSTANT DIAGONALS
20   REM- Z DENOTES OLD SOLUTION , X DENOTES NEW SOLUTION.
40   DIM X(20),B(20),Z(20)
50   PRINT "NUMBER OF EQUATIONS ";
60   INPUT N
70   PRINT "INPUT THE 3 DIAGONAL ENTRIES (LEFT TO RT)"
80   INPUT U,V,W
90   PRINT "INPUT B"
100  MAT INPUT B(N)
110  REM- INITIALIZE Z AS ALL ONES
120  FOR I=1 TO N
130  Z(I)=1
140  NEXT I
150  PRINT "ABSOLUTE ACCURACY REQUIRED ";
160  INPUT E
170  PRINT "MAXIMUM NUMBER OF ITERATIONS REQUIRED ";
180  INPUT M
190  PRINT "DO YOU WISH TO CHOOSE AN INITIAL VECTOR X ?"
200  PRINT "TYPE 0 FOR NO , TYPE 1 FOR YES"
210  INPUT C
```

```
220 IF C=0 THEN 250
230 PRINT "INPUT INITIAL VECTOR X"
240 MAT INPUT Z(N)
250 K=0
260 REM- ITERATION (K) STARTS HERE
270 REM- CALCULATE X(I) FROM Z(I)
280 REM- CURRENT ACCURACY =D = MAX ABS (X(I)-Z(I))
290 K=K+1
300 D=0
310 PRINT
320 PRINT "ITERATION",K
330 FOR I=1 TO N
340 C=B(I)
350 IF I=1 THEN 390
370 C=C-U*Z(I-1)
390 IF I=N THEN 430
410 C=C-W*Z(I+1)
430 X(I)=C/V
440 F=ABS(X(I)-Z(I))
450 IF D>F THEN 470
460 D=F
470 NEXT I
480 PRINT "VECTOR X :"
490 FOR I=1 TO N
500 Z(I)=X(I)
510 PRINT X(I),
520 NEXT I
530 PRINT
540 REM- TEST ACCURACY AND NUMBER OF ITERATIONS
550 IF D<=E THEN 580
560 IF K>=M THEN 600
570 GO TO 290
580 PRINT "ABSOLUTE ACCURACY ATTAINED"
590 GO TO 610
600 PRINT "MAX NO OF ITERATIONS REACHED"
610 END

Ready
```

## Sample run

```
RUN
JACTRI

NUMBER OF EQUATIONS ? 4
INPUT THE 3 DIAGONAL ENTRIES (LEFT TO RT)
? 1,-10,1
INPUT B
? 1,2,3,4
ABSOLUTE ACCURACY REQUIRED ? .001
MAXIMUM NUMBER OF ITERATIONS REQUIRED ? 10
DO YOU WISH TO CHOOSE AN INITIAL VECTOR X ?
TYPE 0 FOR NO , TYPE 1 FOR YES
? 0

ITERATION       1
VECTOR X :
 0              0             -.1            -.3

ITERATION       2
VECTOR X :
-.1            -.21           -.33           -.41

ITERATION       3
VECTOR X :
-.121          -.243          -.362          -.433

ITERATION       4
VECTOR X :
-.1243         -.2483         -.3676         -.4362

ITERATION       5
VECTOR X :
-.12483        -.24919        -.36845        -.43676
ABSOLUTE ACCURACY ATTAINED
Ready
```

*Program notes*

(1) U, V, W are exactly the constants $u$, $v$, $w$ of (6.10).

(2) This Program is formed by making the following changes to Program 6.1:

10  REM– JACTRI: SOLVES AX = B BY JACOBI ITERATION

15  REM– A IS TRIDIAGONAL WITH 3 CONSTANT DIAGONALS

70  PRINT "INPUT THE 3 DIAGONAL ENTRIES (LEFT TO RT)"

80  INPUT U, V, W

370  C = C – U* Z(I – 1)

410  C = C – W* Z(I + 1)

430  X(I) = C/V

DELETE 30, 360, 380, 400, 420

(3) In Sample run 1 we solve for $x_1$, $x_2$, $x_3$, $x_4$ the system

$$x_{i-1} - 10x_i + x_{i+1} = i \quad (i = 1, 2, 3, 4)$$

where $x_0 = x_5 = 0$

i.e.

$$\begin{bmatrix} -10 & 1 & 0 & 0 \\ 1 & -10 & 1 & 0 \\ 0 & 1 & -10 & 1 \\ 0 & 0 & 1 & -10 \end{bmatrix} \begin{bmatrix} x_1 \\ x_2 \\ x_3 \\ x_4 \end{bmatrix} = \begin{bmatrix} 1 \\ 2 \\ 3 \\ 4 \end{bmatrix}$$

The correctness of the computed solution may easily be checked. [This problem is in fact based on a finite difference method for determining the values $x_1, x_2, x_3, x_4$ at $t = .2, .4, .6, .8$ of the solution $x(t)$ of the differential equation

$$\frac{d^2x}{dt^2} - 200x = 125t \text{ for } x(0) = x(1) = 0.]$$

(4) This program may easily be generalised to solve 'multidiagonal' systems with any number of diagonals of entries (see Problem 6).

### 6.2.2 *Convergence*

Jacobi's iteration does *not* necessarily converge to a solution of $\mathbf{Ax} = \mathbf{b}$. However, it is not difficult to deduce one situation in which convergence is assured, and that occurs when

$$|a_{ii}| > \sum_{j=1}^{i-1} |a_{ij}| + \sum_{j=i+1}^{n} |a_{ij}| = \sum_{\substack{j=1 \\ j \neq i}}^{n} |a_{ij}| \qquad (6.11)$$

The latter condition defines a matrix **A** which is *strictly diagonally dominant*, in which the magnitude of any term on the diagonal exceeds the sum of the magnitudes of all terms off the diagonal in the same row. Thus for matrix (6.10), for example, this corresponds to the condition:

$$|v| > |u| + |w|$$

The proof of the sufficiency of condition (6.11) for convergence is straightforward. For equations (6.7) imply that the true solution $x_1, \ldots, x_n$ satisfies

$$x_i = \left( b_i - \sum_{j=1}^{i-1} a_{ij}x_j - \sum_{j=i+1}^{n} a_{ij}x_j \right) / a_{ii}$$

Combining this equation with (6.9), we obtain

$$x_i - x_i^{(k)} = -(a_{ii})^{-1} \left[ \sum_{j=1}^{i-1} a_{ij}(x_j - x_j^{(k-1)}) + \sum_{j=i+1}^{n} a_{ij}(x_j - x_j^{(k-1)}) \right]$$

Thus $\qquad e_i^{(k)} = -(a_{ii})^{-1} \left[ \sum_{j=1}^{i-1} a_{ij}e_j^{(k-1)} + \sum_{j=i+1}^{n} a_{ij}e_j^{(k-1)} \right]$

where $e_i^{(k)} = x_i - x_i^{(k)} =$ truncation error in $x_i$ at iteration $(k)$.

Hence $\qquad |e_i^{(k)}| \leqslant L. \max_{i=1,\ldots,n} |e_i^{(k-1)}|$

where $\qquad L = \max_i \left[ |a_{ii}|^{-1} . \left( \sum_{j=1}^{i-1} |a_{ij}| + \sum_{j=i+1}^{n} |a_{ij}| \right) \right] \qquad (6.12)$

Taking the maximum over $i$ of the left-hand side:

$$\max_i |e_i^{(k)}| \leqslant L.\max_i |e_i^{(k-1)}| \qquad (6.13)$$

Now $\max_i |e_i^{(k)}|$ represents the largest error in the components $x_i^{(k)}$ of $\mathbf{x}^{(k)}$ in iteration $(k)$, and so it measures the *overall error* in $\mathbf{x}^{(k)}$. Also the constant $L$ given by (6.12) may be calculated from **A**, and it is always less than unity as a consequence of assumption (6.11). Indeed $L$ is a *measure of diagonal dominance* since it represents the ratio of the off-diagonal terms to the diagonal term. Now the inequality (6.13) expresses the fact that the overall error is decreasing at (at least) a constant rate $L$ at each step. We therefore say that the Jacobi iteration is *linearly convergent*, provided that **A** is strictly diagonally dominant. (See Chapter 4 of Reference 1 for a discussion of linear convergence in the solution of a nonlinear algebraic equation.)

## 6.3  Iteration method of Gauss–Seidel

By making a very small change in Jacobi's method we can generally improve its convergence. In formula (6.8) which defines $x_1^{(1)}, \ldots, x_n^{(1)}$ from $x_1^{(0)}, \ldots, x_n^{(0)}$, we may update each value of $x_1^{(0)}, \ldots, x_n^{(0)}$ on the right-hand side to $x_1^{(1)}, \ldots, x_n^{(1)}$ as soon as any of the latter values have been calculated. The amended version becomes:

$$a_{11}x_1^{(1)} = b_1 - a_{12}x_2^{(0)} - a_{13}x_3^{(0)} - \ldots - a_{1n}x_n^{(0)}$$
$$a_{22}x_2^{(1)} = b_2 - a_{21}x_1^{(1)} - a_{23}x_3^{(0)} - \ldots - a_{2n}x_n^{(0)}$$
$$a_{33}x_3^{(1)} = b_3 - a_{31}x_1^{(1)} - a_{32}x_2^{(1)} - a_{34}x_4^{(0)} - \ldots - a_{3n}x_n^{(0)}$$
$$\vdots$$
$$a_{nn}x_n^{(1)} = b_n - a_{n1}x_1^{(1)} - a_{n2}x_2^{(1)} - \ldots a_{n,n-1}x_{n-1}^{(1)}$$

For example, equation (6.5) for problem (6.2) would be amended to

$$-30x_1^{(1)} = -30 - 0 \quad -1 \quad -0 \quad = -31$$
$$\Rightarrow x_1^{(1)} = 1.033$$
$$-30x_2^{(1)} = \quad 0 - 1.033 - 1 \quad -0 \quad = -\ 2.033$$
$$\Rightarrow x_2^{(1)} = \ .068$$
$$-30x_3^{(1)} = -30 - 1.033 - \ .068 - 0 \quad = -31.101$$
$$\Rightarrow x_3^{(1)} = 1.037$$
$$-30x_4^{(1)} = \quad 0 - 1.033 - \ .068 - 1.037 = -\ 2.138$$
$$\Rightarrow x_4^{(1)} = \ .071$$

The resulting values $x_1^{(1)}, x_2^{(1)}, x_3^{(1)}, x_4^{(1)}$ are seen to be progressively more accurate, indeed $x_4^{(1)}$ is correct to 3 decimals while $x_1^{(1)}$ is only correct to 2 decimals. Indeed it can be proved that the Gauss–Seidel method generally converges about twice as fast as the Jacobi method (see Reference 4 for details).

The formal algorithm is as follows:

*Algorithm 6.3  Gauss–Seidel iteration*

Perform Algorithm 6.1 with the following modification:

(ii) $x_i^{(k)} = \left( b_i - \sum_{j=1}^{i-1} a_{ij}x_j^{(k)} - \sum_{j=i+1}^{n} a_{ij}x_j^{(k-1)} \right) / a_{ii}$

The Gauss–Seidel method is usually preferable to the Jacobi method from the point of view of computer implementation, since its convergence is generally superior while its arithmetic is no more expensive. However, for hand computation, the simplicity of the Jacobi method still makes it an attractive method to use.

As far as rigorous convergence results are concerned there are two important cases which should be mentioned. Firstly, the Gauss–Seidel method always converges if **A** is strictly diagonally dominant. This is a type of matrix we have considered already in connection with Jacobi's method. Secondly, it always converges if **A** is symmetric and positive definite (see Reference 3 for a proof).

A simple example of a positive definite symmetric matrix is the normal matrix $\mathbf{A} = \mathbf{B}^{\mathrm{T}}\mathbf{B}$, occurring in the application of the least squares method to an over-determined system of $m$ equations in $n$ unknowns $\mathbf{B}\mathbf{x} = \mathbf{y}$ (see Chapter 3). So the Gauss–Seidel method could in principle be used to solve such problems (see Chapter 7).

The method is now coded in Program 6.3.

**Program 6.3** GAUSEI: Gauss–Seidel iteration

```
LIST
GAUSEI

10   REM- GAUSEI: SOLVES AX=B BY GAUSS-SEIDEL ITERATION.
20   REM- Z DENOTES OLD SOLUTION , X DENOTES NEW SOLUTION.
30   DIM A(20,20)
40   DIM X(20),B(20),Z(20)
50   PRINT "NUMBER OF EQUATIONS ";
60   INPUT N
70   PRINT "INPUT A ROW BY ROW "
80   MAT INPUT A(N,N)
90   PRINT "INPUT B"
100  MAT INPUT B(N)
110  REM- INITIALIZE Z AS ALL ONES
120  FOR I=1 TO N
130  Z(I)=1
140  NEXT I
150  PRINT "ABSOLUTE ACCURACY REQUIRED ";
160  INPUT E
170  PRINT "MAXIMUM NUMBER OF ITERATIONS REQUIRED ";
180  INPUT M
190  PRINT "DO YOU WISH TO CHOOSE AN INITIAL VECTOR X ?"
200  PRINT "TYPE 0 FOR NO , TYPE 1 FOR YES"
210  INPUT C
220  IF C=0 THEN 250
230  PRINT "INPUT INITIAL VECTOR X"
240  MAT INPUT Z(N)
250  K=0
260  REM- ITERATION (K) STARTS HERE
270  REM- CALCULATE X(I) FROM Z(I)
280  REM- CURRENT ACCURACY =D = MAX ABS (X(I)-Z(I))
290  K=K+1
300  D=0
310  PRINT
320  PRINT "ITERATION",K
330  FOR I=1 TO N
340  C=B(I)
350  IF I=1 THEN 390
360  FOR J=1 TO I-1
370  C=C-A(I,J)*X(J)
380  NEXT J
390  IF I=N THEN 430
400  FOR J=I+1 TO N
410  C=C-A(I,J)*Z(J)
420  NEXT J
430  X(I)=C/A(I,I)
440  F=ABS(X(I)-Z(I))
450  IF D>F THEN 470
460  D=F
470  NEXT I
480  PRINT "VECTOR X :"
490  FOR I=1 TO N
500  Z(I)=X(I)
510  PRINT X(I),
```

```
520 NEXT I
530 PRINT
540 REM- TEST ACCURACY AND NUMBER OF ITERATIONS
550 IF D<=E THEN 580
560 IF K>=M THEN 600
570 GO TO 290
580 PRINT "ABSOLUTE ACCURACY ATTAINED"
590 GO TO 610
600 PRINT "MAX NO OF ITERATIONS REACHED"
610 END
Ready
```

## Sample run 1

```
RUN
GAUSEI

NUMBER OF EQUATIONS ? 4
INPUT A ROW BY ROW
? -30,1,1,1,-30,1,1,1,1,-30,1,1,1,1,-30
INPUT B
? -30,0,-30,0
ABSOLUTE ACCURACY REQUIRED ? .00001
MAXIMUM NUMBER OF ITERATIONS REQUIRED ? 10
DO YOU WISH TO CHOOSE AN INITIAL VECTOR X ?
TYPE 0 FOR NO , TYPE 1 FOR YES
? 1
INPUT INITIAL VECTOR X
? 1,0,1,0

ITERATION       1
VECTOR X :
 1.03333        .677778E-01    1.0367         .712605E-01

ITERATION       2
VECTOR X :
 1.03919        .715718E-01    1.0394         .716721E-01

ITERATION       3
VECTOR X :
 1.03942        .716832E-01    1.03943        .716844E-01

ITERATION       4
VECTOR X :
 1.03943        .716846E-01    1.03943        .716846E-01
ABSOLUTE ACCURACY ATTAINED
Ready
```

## Sample run 2

```
RUN
GAUSEI

NUMBER OF EQUATIONS ? 3
INPUT A ROW BY ROW
? 2,1,4,    1,2,4,    4,4,11
INPUT B
? 7,7,19
ABSOLUTE ACCURACY REQUIRED ? .001
MAXIMUM NUMBER OF ITERATIONS REQUIRED ? 10
DO YOU WISH TO CHOOSE AN INITIAL VECTOR X ?
TYPE 0 FOR NO , TYPE 1 FOR YES
? 1
INPUT INITIAL VECTOR X
? .9,.9,.9

ITERATION       1
VECTOR X :
 1.25           1.075          .881818

ITERATION       2
VECTOR X :
 1.19886        1.13693        .877892

ITERATION       3
VECTOR X :
 1.17575        1.15634        .87924
```

```
ITERATION        4
VECTOR X :
  1.16335        1.15984        .882475

ITERATION        5
VECTOR X :
  1.15513        1.15749        .886322

ITERATION        6
VECTOR X :
  1.14861        1.15305        .890305

ITERATION        7
VECTOR X :
  1.14287        1.14796        .894246

ITERATION        8
VECTOR X :
  1.13753        1.14274        .898083

ITERATION        9
VECTOR X :
  1.13246        1.1376         .901794

ITERATION       10
VECTOR X :
  1.12761        1.13261        .905376
MAX NO OF ITERATIONS REACHED
Ready
```

*Program notes*

(1) The program is identical to Program 6.1 apart from the following two changes:

    10 REM–GAUSEI: SOLVES AX = B BY GAUSS–SEIDEL
           ITERATION
    370 C = C − A(I, J)*X(J)

(2) In Sample run 1, the program succeeds in solving the strictly diagonally dominant system (6.2), and it can be seen (by comparison with Sample run 1 of Program 6.1) to require fewer iterations than Jacobi's method.

(3) In Sample run 2, the program tackles a system in which $\mathbf{A}$ is the positive definite symmetric matrix $\mathbf{A} = \mathbf{B}^{\mathsf{T}}\mathbf{B}$, where

$$\mathbf{A} = \begin{bmatrix} 2 & 1 & 4 \\ 1 & 2 & 4 \\ 4 & 4 & 11 \end{bmatrix}, \ \mathbf{B} = \begin{bmatrix} 1 & 0 & 1 \\ 0 & 1 & 1 \\ 1 & 1 & 3 \end{bmatrix}$$

(Note that Jacobi's method *failed* for this system in Sample run 2 of the Program 6.1.) However, although the Gauss–Seidel method converges, it does so *extremely slowly*. Indeed the errors in all components of $x$ are still about .04 after 40 iterations.

Note also that the *error criterion is not meaningful* for such slow convergence. For the maximum change in the components is less than 0.01 after 5 iterations, but the components themselves are in error by more than 0.1.

## 6.4 Successive over-relaxation (S.O.R.)

Iteration methods of the type that we have just described, in which values of the solution are repeatedly updated, are sometimes called *relaxation methods*. It has been found in practice that the correction made to the solution in any iteration $(k)$ of the Gauss–Seidel method, namely

$$\delta \mathbf{x}^{(k)} = \mathbf{x}^{(k)} - \mathbf{x}^{(k-1)} \qquad (6.14)$$

is often an underestimate of the change $\mathbf{x} - \mathbf{x}^{(k-1)}$ which should ideally be made. It is therefore generally accepted that, instead of determining $\mathbf{x}^{(k)}$ as

$$\mathbf{x}^{(k)} = \mathbf{x}^{(k-1)} + \delta \mathbf{x}^{(k)}$$

where $\delta \mathbf{x}^{(k)}$ is the correction adopted in the Gauss–Seidel iteration, $\mathbf{x}^{(k)}$ should be determined by over-correction, or *over-relaxation*, as

$$\mathbf{x}^{(k)} = \mathbf{x}^{(k-1)} + \omega \delta \mathbf{x}^{(k)} \qquad (6.15)$$

where $\omega$ $(>1)$ is a *relaxation factor*. Typically $\omega$ is fixed for all $k$ at a value suitably chosen somewhere between 1 and 2. The iteration (6.15) is called Successive Over-Relaxation, or S.O.R.

In some cases it is possible to choose an 'optimal' relaxation factor $\omega$, based on knowledge of the eigenvalues of $\mathbf{A}$, and an appropriate formula and some numerical examples are to be found in Reference 4. For optimal values of $\omega$, the rate of convergence of the method can be as much as 10 to 100 times as fast as that of Gauss–Seidel.

An algorithm to perform S.O.R. simply involves calculating the Gauss–Seidel vector $\mathbf{x}^{(k)}$, determining $\delta \mathbf{x}^{(k)}$ from (6.14), and modifying $\mathbf{x}^{(k)}$ according to (6.15). The algorithm and program are therefore as follows (with $\mathbf{y}^{(k)}$ denoting the intermediate Gauss–Seidel vector $\mathbf{x}^{(k)}$). Note that Algorithm 6.4 includes Algorithm 6.3 (Gauss–Seidel) as a special case for the choice $\omega = 1$.

*Algorithm 6.4 Successive over-relaxation*

(1) Choose initial values $x_1^{(0)}, \ldots, x_n^{(0)}$. If no choice is obvious, take $x_1^{(0)} = \ldots = x_n^{(0)} = 1$.

(2) Choose a relaxation factor $\omega : 1 < \omega < 2$.

For $k = 1, 2, \ldots$:

(3) Calculate $y_i^{(k)} = \left( b_i - \sum_{j=1}^{i-1} a_{ij} x_j^{(k)} - \sum_{j=i+1}^{n} a_{ij} x_j^{(k-1)} \right) \Big/ a_{ii}$

$$(i = 1, \ldots, n)$$

(4) Calculate $x_i^{(k)} = x_i^{(k-1)} + \omega ( y_i^{(k)} - x_i^{(k-1)} )$   $(i = 1, \ldots, n)$

(5)  Terminate iteration on $k$ when $|x_i^{(k)} - x_i^{(k-1)}|$ is sufficiently small for every $i$.

**Program 6.4** S.O.R.: Successive Over-relaxation

```
LIST
SOR

10   REM- SOR: SOLVES AX=B BY SUCCESSIVE OVER-RELAXATION
20   REM- Z,Y,X DENOTE OLD, GAUSS-SEIDEL, AND NEW SOLUTIONS
30   DIM A(20,20)
40   DIM X(20),B(20),Z(20)
50   PRINT "NUMBER OF EQUATIONS ";
60   INPUT N
70   PRINT "INPUT A ROW BY ROW "
80   MAT INPUT A(N,N)
90   PRINT "INPUT B"
100  MAT INPUT B(N)
101  GO TO 110
102  PRINT "IF YOU WANT ANOTHER W TYPE 1, OTHERWISE 0";
103  INPUT G
104  IF G=0 THEN 620
110  REM- INITIALIZE Z AS ALL ONES
120  FOR I=1 TO N
130  Z(I)=1
140  NEXT I
142  PRINT "INPUT RELAXATION FACTOR W ";
144  INPUT W
150  PRINT "ABSOLUTE ACCURACY REQUIRED ";
160  INPUT E
170  PRINT "MAXIMUM NUMBER OF ITERATIONS REQUIRED ";
180  INPUT M
190  PRINT "DO YOU WISH TO CHOOSE AN INITIAL VECTOR X ?"
200  PRINT "TYPE 0 FOR NO , TYPE 1 FOR YES"
210  INPUT C
220  IF C=0 THEN 250
230  PRINT "INPUT INITIAL VECTOR X"
240  MAT INPUT Z(N)
250  K=0
260  REM- ITERATION (K) STARTS HERE
270  REM- CALCULATE X(I) FROM Z(I)
280  REM- CURRENT ACCURACY =D = MAX ABS (X(I)-Z(I))
290  K=K+1
300  D=0
320  PRINT "ITERATION",K
330  FOR I=1 TO N
340  C=B(I)
350  IF I=1 THEN 390
360  FOR J=1 TO I-1
370  C=C-A(I,J)*X(J)
380  NEXT J
390  IF I=N THEN 430
400  FOR J=I+1 TO N
410  C=C-A(I,J)*Z(J)
420  NEXT J
430  Y=C/A(I,I)
435  X(I)=Z(I)+W*(Y-Z(I))
440  F=ABS(X(I)-Z(I))
450  IF D>F THEN 470
460  D=F
470  NEXT I
480  PRINT "VECTOR X :"
490  FOR I=1 TO N
500  Z(I)=X(I)
510  PRINT X(I),
520  NEXT I
530  PRINT
540  REM- TEST ACCURACY AND NUMBER OF ITERATIONS
550  IF D<=E THEN 580
560  IF K>=M THEN 600
570  GO TO 290
580  PRINT "ABSOLUTE ACCURACY ATTAINED"
590  GO TO 610
600  PRINT "MAX NO OF ITERATIONS REACHED"
610  GO TO 102
620  END

Ready
```

## Sample run 1

```
RUN
SOR

NUMBER OF EQUATIONS ? 4
INPUT A ROW BY ROW
? 4,0,1,1,  0,4,0,1,  1,0,4,0,  1,1,0,4
INPUT B
? 1,2,3,4
INPUT RELAXATION FACTOR W ? 1
ABSOLUTE ACCURACY REQUIRED ? .00001
MAXIMUM NUMBER OF ITERATIONS REQUIRED ? 20
DO YOU WISH TO CHOOSE AN INITIAL VECTOR X ?
TYPE 0 FOR NO , TYPE 1 FOR YES
? 0
ITERATION     1
VECTOR X :
-.25          .25           .8125         1
ITERATION     2
VECTOR X :
-.203125      .25           .800781       .988281
ITERATION     3
VECTOR X :
-.197266      .25293        .799316       .986084
ITERATION     4
VECTOR X :
-.19635       .253479       .799088       .985718
ITERATION     5
VECTOR X :
-.196201      .253571       .79905        .985658
ITERATION     6
VECTOR X :
-.196177      .253586       .799044       .985648
ITERATION     7
VECTOR X :
-.196173      .253588       .799043       .985646
ABSOLUTE ACCURACY ATTAINED
IF YOU WANT ANOTHER W TYPE 1, OTHERWISE 0? 1
INPUT RELAXATION FACTOR W ? 1.05
ABSOLUTE ACCURACY REQUIRED ? .00001
MAXIMUM NUMBER OF ITERATIONS REQUIRED ? 20
DO YOU WISH TO CHOOSE AN INITIAL VECTOR X ?
TYPE 0 FOR NO , TYPE 1 FOR YES
? 0
ITERATION     1
VECTOR X :
-.3125        .2125         .819531       1.02625
ITERATION     2
VECTOR X :
-.206393      .244984       .800701       .988557
ITERATION     3
VECTOR X :
-.196861      .253255       .799141       .985769
ITERATION     4
VECTOR X :
-.196196      .253573       .799044       .98565
ITERATION     5
VECTOR X :
-.196172      .253588       .799043       .985646
ITERATION     6
VECTOR X :
-.196172      .253589       .799043       .985646
ABSOLUTE ACCURACY ATTAINED
IF YOU WANT ANOTHER W TYPE 1, OTHERWISE 0? 0
Ready
```

## Sample run 2

```
RUN
SOR

NUMBER OF EQUATIONS ? 3
INPUT A ROW BY ROW
? 2,1,4,   1,2,4,   4,4,11
INPUT B
? 7,7,19
```

```
INPUT RELAXATION FACTOR W ? 1.6
ABSOLUTE ACCURACY REQUIRED ? .001
MAXIMUM NUMBER OF ITERATIONS REQUIRED ? 10
DO YOU WISH TO CHOOSE AN INITIAL VECTOR X ?
TYPE 0 FOR NO , TYPE 1 FOR YES
? 1
INPUT INITIAL VECTOR X
? .9,.9,.9
ITERATION        1
VECTOR X :
  1.46           1.012          .785382
ITERATION        2
VECTOR X :
  1.40118        1.35864        .686697
ITERATION        3
VECTOR X :
  1.47495        1.40742        .674598
ITERATION        4
VECTOR X :
  1.43037        1.45253        .68155
ITERATION        5
VECTOR X :
  1.39879        1.42849        .709745
ITERATION        6
VECTOR X :
  1.34675        1.39432        .742983
ITERATION        7
VECTOR X :
  1.29895        1.3467         .778559
ITERATION        8
VECTOR X :
  1.25188        1.29908        .812304
ITERATION        9
VECTOR X :
  1.21023        1.25299        .843106
ITERATION        10
VECTOR X :
  1.17353        1.21144        .870152
MAX NO OF ITERATIONS REACHED
IF YOU WANT ANOTHER W TYPE 1, OTHERWISE 0? 0
Ready
```

## Program notes

(1) This program is formed by making the following changes to Program 6.3. (The numbering of instructions used in Program 6.3 is retained.) New REM statements 10 and 20 are introduced for explanatory purposes. Instructions 101 to 104, 610 allow a transfer of control to instruction 110 at the end of the calculation, so that the calculation can be repeated with a new relaxation factor $\omega$ (i.e. W) if required. Instructions 142 and 144 input this factor W. Finally instructions 430, 435 change the Gauss–Seidel iteration into the S.O.R. method.

(2) The Gauss–Seidel vector **y** has components $y_1$, $y_2$, ..., $y_n$. However, only one location Y is needed in the program for all $y_i$, since each component is only calculated temporarily.

(3) In the first Sample run, the program solves a strictly diagonally dominant system successfully, and it can be seen that convergence is faster for the choice $\omega = 1.05$ than for the choice $\omega = 1$ (i.e. Gauss–Seidel).

(4) In the second Sample run, a system with a positive definite symmetric matrix is tackled. This is the same example as was tackled

by Gauss–Seidel iteration in Sample run 2 of Program 6.3. Although these results may not look promising after 10 iterations, the choice $\omega = 1.6$ is in fact nearly optimal. Indeed after 29 iterations the components in $\mathbf{x}$ may be found to be changing by at most 0.0001 and to be correct to within 0.004. (After 29 iterations, the Gauss–Seidel vector components are only correct to within 0.07.)

## 6.5  Eigenvalues and eigenvectors of a matrix

An important matrix problem, called the 'eigenvalue problem', is to determine a constant $\lambda$ and a corresponding non-zero vector $\mathbf{x}$ such that

$$\mathbf{Ax} = \lambda\mathbf{x} \qquad (6.16)$$

where $\mathbf{A}$ is a given square matrix. (The constant $\lambda$ is called an eigenvalue, and the vector $\mathbf{x}$ is called an eigenvector.) If $\mathbf{A}$ is an $n$ by $n$ matrix then $\mathbf{x} = (x_1, x_2, \ldots, x_n)^{\mathrm{T}}$ is an $n$ by 1 matrix. By a non-zero vector $\mathbf{x}$, we mean one which does not have all its components $x_1$, $x_2, \ldots, x_n$ equal to zero. There are in general $n$ independent solution pairs $\lambda$, $\mathbf{x}$ of (6.16), and we may require one solution, a few solutions, or all solutions. The eigenvalue problem occurs in many practical applications, such as the determination of principal stress directions of a material and principal axes of inertia, the calculation of displacements of a body under structural vibrations, the solutions of coupled linear differential equations, and so on.

An eigenvector is only unique up to a multiplicative constant. In other words, as can immediately be seen from (6.16), if $\mathbf{x}$ is an eigenvector then so also is $c\mathbf{x}$ for any non-zero constant $c$. For this reason $\mathbf{x}$ is usually thought of as representing a *direction* in space (rather than a point in space) defined by the line joining the origin $(0, 0, \ldots, 0)$ to the point $(x_1, x_2, \ldots, x_n)$. It is then obvious that the zero vector $(0\ 0 \ldots 0)^{\mathrm{T}}$, which does not represent a direction, is not normally acceptable as an eigenvector.

A 'practical mathematical' way of interpreting the relation (6.16) is as follows. A matrix $\mathbf{A}$ can represent a *linear transformation* of variables

$$\mathbf{y} = \mathbf{Ax} \qquad (6.17)$$

from the variable $\mathbf{x}$ to the variable $\mathbf{y}$. (See Section 3.5.) For example, in three dimensions (6.17) gives

$$\left.\begin{array}{l} y_1 = a_{11}x_1 + a_{12}x_2 + a_{13}x_3 \\ y_2 = a_{21}x_1 + a_{22}x_2 + a_{23}x_3 \\ y_3 = a_{31}x_1 + a_{32}x_2 + a_{33}x_3 \end{array}\right\} \qquad (6.18)$$

and each direction $(x_1\ x_2\ x_3)^T$ in space is transformed into a new direction $(y_1\ y_2\ y_3)^T$ by the matrix $\mathbf{A}$. If the new direction happens to coincide with the original direction then

$$\mathbf{y} = \lambda\mathbf{x} \quad \text{(for some constant } \lambda)$$

and hence from (6.17)

$$\mathbf{A}x = \lambda\mathbf{x}$$

Thus an eigenvector $\mathbf{x}$ is a direction which is *invariant* (i.e. unchanged) *under a linear transformation* of variables $\mathbf{A}$; and the corresponding eigenvalue $\lambda$ measures the 'expansion' (from $\mathbf{x}$ to $\lambda\mathbf{x}$) of the eigenvector under the transformation.

### 6.5.1 *Example of eigenvalues*

A simple example of an eigenvalue problem is

$$\mathbf{A}x = \lambda\mathbf{x} \text{ for } \mathbf{A} = \begin{bmatrix} 99 & 2 \\ 49 & 2 \end{bmatrix} \text{ and } \mathbf{x} = \begin{bmatrix} x_1 \\ x_2 \end{bmatrix}$$

The classical mathematical approach is as follows.

Clearly

$$\mathbf{A}x - \lambda\mathbf{x} = (\mathbf{A} - \lambda\mathbf{I})\mathbf{x} = \begin{bmatrix} 99-\lambda & 2 \\ 49 & 2-\lambda \end{bmatrix} \begin{bmatrix} x_1 \\ x_2 \end{bmatrix} = \mathbf{0}$$

For a valid solution of this pair of homogeneous equations, it is necessary that the determinant of $\mathbf{A} - \lambda\mathbf{I}$ should vanish, and hence

$$|\mathbf{A} - \lambda\mathbf{I}| = \begin{vmatrix} 99-\lambda & 2 \\ 49 & 2-\lambda \end{vmatrix} = 0 \tag{6.19}$$

Now (6.19) may be expanded to give the quadratic equation

$$0 = (99-\lambda)(2-\lambda) - 98 = 100 - 101\lambda + \lambda^2$$

and hence $\qquad \lambda = 100 \text{ or } 1$

For $\lambda = 100$, the equation for the corresponding eigenvector $\mathbf{x}$ is

$$(\mathbf{A} - \lambda\mathbf{I})\mathbf{x} = \begin{bmatrix} -1 & 2 \\ 49 & -98 \end{bmatrix} \begin{bmatrix} x_1 \\ x_2 \end{bmatrix} = \begin{bmatrix} 0 \\ 0 \end{bmatrix} \tag{6.20}$$

and we deduce that $\qquad \mathbf{x} = (2\ 1)^T$

For $\lambda = 1$, the corresponding eigenvector $\mathbf{x}$ is given by

$$(\mathbf{A} - \lambda\mathbf{I})\mathbf{x} = \begin{bmatrix} 98 & 2 \\ 49 & 1 \end{bmatrix} \begin{bmatrix} x_1 \\ x_2 \end{bmatrix} = \begin{bmatrix} 0 \\ 0 \end{bmatrix}$$

and we deduce that $\qquad \mathbf{x} = (1 - 49)^{\mathrm{T}}$

Thus there are two solution pairs

$$\lambda = 100, \mathbf{x} = (2\ 1)^{\mathrm{T}}; \quad \lambda = 1, \mathbf{x} = (1 - 49)^{\mathrm{T}}$$

### 6.5.2 Numerical approach

The numerical mathematical approach to the eigenvalue problem is for the most part quite different from the classical approach above, although we shall return to (6.19) and (6.20) later in the context of general algorithms. For simplicity we shall only consider the case, which is common to many practical problems, in which all eigenvalues and eigenvectors are real, distinct, and non-zero. We shall also assume that the set of eigenvectors is 'linearly independent', and this means that one eigenvector, $\mathbf{x}_{(1)}$ say, *cannot* be expressed as a linear combination $c_2\mathbf{x}_{(2)} + \ldots + c_n\mathbf{x}_{(n)}$ of the other eigenvectors $\mathbf{x}_{(2)}, \ldots, \mathbf{x}_{(n)}$.

We shall start by discussing simple algorithms for determining particular eigenvalues (such as the largest, smallest, etc.) and shall leave until last a discussion of general algorithms.

### 6.6 The power method for the largest eigenvalue

Suppose that $\mathbf{A}$ (of order $n$) has $n$ distinct real eigenvalues $\lambda_1, \lambda_2, \ldots, \lambda_n$ and $n$ corresponding linearly independent eigenvectors $\mathbf{x}_{(1)}, \mathbf{x}_{(2)}, \ldots, \mathbf{x}_{(n)}$. Suppose also that $\lambda_1, \lambda_2, \ldots, \lambda_n$ are in decreasing order of magnitude, so that

$$|\lambda_1| > |\lambda_2| > \ldots > |\lambda_n| \qquad (6.21)$$

(This implies for example that no two eigenvalues have equal magnitudes and opposite signs.) Then the 'power method' for determining the largest eigenvalue $\lambda_1$ and corresponding eigenvector $\mathbf{x}_{(1)}$ is based on the following simple idea. Calculate a sequence of vectors $\mathbf{y}^{(1)}, \mathbf{y}^{(2)}, \ldots, \mathbf{y}^{(k)}, \ldots$, starting from a chosen initial vector $\mathbf{y}^{(0)}$, by defining $\mathbf{y}^{(k)}$ from $\mathbf{y}^{(k-1)}$ by the iteration

$$\mathbf{y}^{(k)} = \mathbf{A}\mathbf{y}^{(k-1)} \quad (k = 1, 2, \ldots) \qquad (6.22)$$

(In other words, start with a vector $\mathbf{y}^{(0)}$ and keep multiplying it by $\mathbf{A}$.) Then, as $k \to \infty$, $\mathbf{y}^{(k)}$ approaches $\mathbf{x}_{(1)}$ (to within a multiplicative constant) and $\mathbf{y}^{(k)}$ approaches a constant $\lambda_1$ times $\mathbf{y}^{(k-1)}$.

### 6.6.1 *Example of the power method*

Consider again $\qquad \mathbf{A} = \begin{bmatrix} 99 & 2 \\ 49 & 2 \end{bmatrix}$ $\qquad\qquad$ (6.23)

Then starting from $\mathbf{y}^{(0)} = \begin{bmatrix} 1 \\ 1 \end{bmatrix}$, we obtain

$$\mathbf{y}^{(1)} = \mathbf{A}\mathbf{y}^{(0)} = \begin{bmatrix} 101 \\ 51 \end{bmatrix}$$

$$\mathbf{y}^{(2)} = \mathbf{A}\mathbf{y}^{(1)} = \begin{bmatrix} 10101 \\ 5051 \end{bmatrix}$$

$$\mathbf{y}^{(3)} = \mathbf{A}\mathbf{y}^{(2)} = \begin{bmatrix} 1010101 \\ 505051 \end{bmatrix} \cdots$$

The ratios of consecutive first elements in this sequence of vectors is

$$\frac{101}{1} = 101, \quad \frac{10101}{101} = 100.0099, \quad \frac{1010101}{10101} = 100.000099, \dots,$$

and it is clear that these are converging to the largest eigenvalue 100. From the calculated values alone we would deduce that, to five figures,

$$\lambda_1 = 100.00, \quad \mathbf{x}_{(1)} = \begin{bmatrix} 1010100 \\ 505050 \end{bmatrix}$$

### 6.6.2 *Theory of the power method*

Why does this method work? Suppose that the initial vector $\mathbf{y}^{(0)}$ were expressed as a linear combination of the eigenvectors in the form

$$\mathbf{y}^{(0)} = c_1\mathbf{x}_{(1)} + c_2\mathbf{x}_{(2)} + \dots + c_n\mathbf{x}_{(n)} \qquad (6.24)$$

where $c_1, \dots, c_n$ are some constants (whose specific values are not required). Such a combination can always be found, and it essentially expresses $\mathbf{y}^{(0)}$ in terms of coordinate directions $\mathbf{x}_{(1)}, \dots, \mathbf{x}_{(n)}$ by choosing appropriate coordinates $c_1, \dots, c_n$. Then, from (6.22) and (6.24),

$$\begin{aligned} \mathbf{y}^{(1)} = \mathbf{A}\mathbf{y}^{(0)} &= c_1\mathbf{A}\mathbf{x}_{(1)} + c_2\mathbf{A}\mathbf{x}_{(2)} + \dots + c_n\mathbf{A}\mathbf{x}_{(n)} \\ &= c_1\lambda_1\mathbf{x}_{(1)} + c_2\lambda_2\mathbf{x}_{(2)} + \dots + c_n\lambda_n\mathbf{x}_{(n)} \end{aligned}$$

since $\lambda_1, \dots, \lambda_n$ are the eigenvalues corresponding to $\mathbf{x}_{(1)}, \dots, \mathbf{x}_{(n)}$. Similarly $\mathbf{y}^{(2)} = \mathbf{A}\mathbf{y}^{(1)} = c_1(\lambda_1)^2\mathbf{x}_{(1)} + c_2(\lambda_2)^2\mathbf{x}_{(2)} + \dots + c_n(\lambda_n)^2\mathbf{x}_{(n)}$ and in general

$$\mathbf{y}^{(k)} = \mathbf{A}\mathbf{y}^{(k-1)} = c_1(\lambda_1)^k\mathbf{x}_{(1)} + c_2(\lambda_2)^k\mathbf{x}_{(2)} + \ldots + c_n(\lambda_n)^k\mathbf{x}_{(n)} \quad (6.25)$$

Since $|\lambda_1|$ is greater than $|\lambda_2|$, $|\lambda_3|$, ..., it follows that for large enough $k$

$$\mathbf{y}^{(k)} \simeq c_1(\lambda_1)^k\mathbf{x}_{(1)} \quad (\text{for } c_1 \neq 0) \quad\quad (6.26)$$

so that, apart from a multiplicative constant, $\mathbf{y}^{(k)}$ is converging to $\mathbf{x}_{(1)}$ as $k \to \infty$. Moreover, from (6.26),

$$\mathbf{y}^{(k)} \simeq \lambda_1\mathbf{y}^{(k-1)}$$

and so the ratio of corresponding elements of $\mathbf{y}^{(k)}$ and $\mathbf{y}^{(k-1)}$ is converging to $\lambda_1$.

### 6.6.3 Modification of the method

Although example (6.23) demonstrates how well this simple algorithm works, it also illustrates a drawback. For the elements of the vector $\mathbf{y}^{(k)}$ become very large as the iteration proceeds. To avoid 'accumulator overflow' (i.e. exceeding the largest number the computer recognises) and also to achieve a neat algorithm, each iteration vector $\mathbf{y}^{(k)}$ should be normalised by dividing it by its element $\alpha_k$ (say) of largest magnitude. Then the normalised vector $\mathbf{z}^{(k)} = (\alpha_k)^{-1}\mathbf{y}^{(k)}$ is multiplied by $\mathbf{A}$ to give $\mathbf{y}^{(k+1)}$, and so on. It then follows that $\alpha_k$ itself converges to $\lambda_1$ and $\mathbf{z}^{(k)}$ converges without multiplicative constants to a uniquely normalised eigenvector $\mathbf{x}_{(1)}$.

Repeating example (6.23) with this normalisation and using five significant figure arithmetic, we obtain:

$$\mathbf{y}^{(0)} = \begin{bmatrix} 1 \\ 1 \end{bmatrix}, \quad \alpha_0 = 1, \quad \mathbf{z}^{(0)} = \begin{bmatrix} 1 \\ 1 \end{bmatrix}$$

$$\mathbf{y}^{(1)} = \mathbf{A}\mathbf{z}^{(0)} = \begin{bmatrix} 101 \\ 51 \end{bmatrix}, \quad \alpha_1 = 101, \quad \mathbf{z}^{(1)} = \begin{bmatrix} 1 \\ .50495 \end{bmatrix}$$

$$\mathbf{y}^{(2)} = \mathbf{A}\mathbf{z}^{(1)} = \begin{bmatrix} 100.01 \\ 50.01 \end{bmatrix}, \quad \alpha_2 = 100.01, \quad \mathbf{z}^{(2)} = \begin{bmatrix} 1 \\ .50005 \end{bmatrix}$$

$$\mathbf{y}^{(3)} = \mathbf{A}\mathbf{z}^{(2)} = \begin{bmatrix} 100.00 \\ 50.000 \end{bmatrix}, \quad \alpha_3 = 100.00, \quad \mathbf{z}^{(3)} = \begin{bmatrix} 1 \\ .50000 \end{bmatrix}$$

From the $\alpha$'s and the $\mathbf{z}$'s we immediately deduce that

$$\lambda_1 = 100.00, \quad \mathbf{x}_{(1)} = \begin{bmatrix} 1 \\ .50000 \end{bmatrix}$$

The algorithm and program for the largest eigenvalue are thus expressed formally as follows:

*Algorithm 6.5  Power method for largest eigenvalue*

(i) Choose an initial vector $\mathbf{y}^{(0)}$. If no choice is obvious, take
$y_1^{(0)} = y_2^{(0)} = \ldots = y_n^{(0)} = 1$.
For $k = 0, 1, 2, \ldots$:

(ii) Determine the largest element $\alpha_k$ (in magnitude) in $\mathbf{y}^{(k)}$

(iii) Calculate a normalised vector $\mathbf{z}^{(k)} = (\alpha_k)^{-1} \mathbf{y}^{(k)}$

(iv) Calculate a new vector $\mathbf{y}^{(k+1)} = \mathbf{A}\mathbf{z}^{(k)}$

(v) Repeat the iteration from (ii) until $\alpha_k$ and *all* components of
$\mathbf{z}^{(k)}$ have converged to a specified accuracy.
Then $\alpha_k \simeq \lambda_1$   (the largest eigenvalue)
and $\mathbf{z}^{(k)} \simeq \mathbf{x}_{(1)}$   (the corresponding eigenvector).

**Program 6.5** ELARGE: Power method for largest eigenvalue

```
LIST
ELARGE

10  REM- ELARGE: FINDS LARGEST EIGENVALUE AND CORRESPONDING EIGENVECTOR
20  REM- OF MATRIX A.   USES POWER METHOD WITH NORMALISED VECTORS.
30  REM- ASSUMES LARGEST EIGENVALUE IN MAGNITUDE IS UNIQUE,SIMPLE,REAL.
40  REM- VECTORS: Y ITERATED, Z NORMALISED, Z1 PREVIOUS VALUE.
50  DIM Y(20),Z(20),Z1(20),A(20,20)
60  PRINT "DIMENSION OF A ";
70  INPUT N
80  PRINT "INPUT MATRIX A BY ROWS"
90  MAT INPUT A(N,N)
100 FOR I=1 TO N
110 Z1(I)=1
120 NEXT I
130 PRINT "ABSOLUTE ACCURACY REQUIRED";
140 INPUT E
150 PRINT "MAX NO OF ITERATIONS ALLOWED";
160 INPUT M
170 PRINT "DO YOU WANT TO CHOOSE YOUR OWN INITIAL VECTOR ?"
180 PRINT "INPUT 0 FOR NO , 1 FOR YES"
190 INPUT G
200 IF G=0 THEN 230
210 PRINT "INPUT INITIAL VECTOR:"
220 MAT INPUT Z1(N)
230 K=1
240 REM- ITERATION ON K BEGINS
250 REM- FORM Y=A*Z1, L=LARGEST ELEMENT, Z=Y/L.
260 D=0
270 L=0
280 FOR I=1 TO N
290 Y(I)=0
300 FOR J=1 TO N
310 Y(I)=A(I,J)*Z1(J)+Y(I)
320 NEXT J
330 F=ABS(Y(I))
340 IF D>F THEN 370
350 D=F
360 L=Y(I)
370 NEXT I
380 L1=L
390 PRINT "ITERATION";K
400 PRINT "EIGENVALUE:";L1
410 PRINT "EIGENVECTOR :"
420 FOR I=1 TO N
430 Z(I)=Y(I)/L
440 PRINT Z(I);
450 NEXT I
460 PRINT
470 REM- STOP IF MAX ABS (Z(I)-Z1(I))<=E  OR  K>=M
480 D=0
490 FOR I=1 TO N
500 F=ABS(Z(I)-Z1(I))
510 IF D>F THEN 530
520 D=F
530 Z1(I)=Z(I)
540 NEXT I
```

```
550 IF D<=E THEN 590
560 IF K>=M THEN 610
570 K=K+1
580 GO TO 260
590 PRINT "ABSOLUTE ACCURACY ACHIEVED"
600 GO TO 620
610 PRINT "MAX NO OF ITERATIONS REACHED"
620 END
```

## Sample run 1

```
RUN
ELARGE

DIMENSION OF A ? 3
INPUT MATRIX A BY ROWS
? 2.5,2.2,0.9,1.1,2.5,1.1,1.1,2.5,2.4
ABSOLUTE ACCURACY REQUIRED? .001
MAX NO OF ITERATIONS ALLOWED? 10
DO YOU WANT TO CHOOSE YOUR OWN INITIAL VECTOR ?
INPUT 0 FOR NO , 1 FOR YES
? 0
ITERATION 1
EIGENVALUE: 6
EIGENVECTOR :
 .933333   .783333   1
ITERATION 2
EIGENVALUE: 5.385
EIGENVECTOR :
 .920458   .758589   1
ITERATION 3
EIGENVALUE: 5.30898
EIGENVECTOR :
 .917322   .755132   1
ITERATION 4
EIGENVALUE: 5.29688
EIGENVECTOR :
 .9165   .754573   1
ABSOLUTE ACCURACY ACHIEVED
Ready
```

## Sample run 2

```
RUN
ELARGE

DIMENSION OF A ? 2
INPUT MATRIX A BY ROWS
? 0,1,  2,0
ABSOLUTE ACCURACY REQUIRED? .01
MAX NO OF ITERATIONS ALLOWED? 4
DO YOU WANT TO CHOOSE YOUR OWN INITIAL VECTOR ?
INPUT 0 FOR NO , 1 FOR YES
? 0
ITERATION 1
EIGENVALUE: 2
EIGENVECTOR :
 .5   1
ITERATION 2
EIGENVALUE: 1
EIGENVECTOR :
 1   1
ITERATION 3
EIGENVALUE: 2
EIGENVECTOR :
 .5   1
ITERATION 4
EIGENVALUE: 1
EIGENVECTOR :
 1   1
MAX NO OF ITERATIONS REACHED
Ready
```

*Program notes*

(1) The program is straightforward, and indeed the main calculation is simply that of forming $y = Az$, in instructions 280–320. Much of the remainder of the program is concerned with normalisation (260, 270, 330–370, 420–450), testing for convergence (480–550), and 'administration'.

In the coding of the normalisation, L denotes $\alpha_k$ (the component Y(I) of largest magnitude) and D is $|L|$, so that L converges to the largest eigenvalue.

(2) In Sample run 1, the program successfully finds the largest eigenvaiue of a matrix.

(3) If the specified matrix **A** does not have a dominant real eigenvalue then the method will not converge. For example, in Sample run 2, where **A** has eigenvalues $\pm\sqrt{2}$, the iteration simply oscillates between two vectors (which are not eigenvectors).

(4) If the initial vector happens to be *exactly* an eigenvector, corresponding to any of the eigenvalues, then obviously the algorithm will not change this vector, regardless of whether or not the eigenvalue is the largest one. In such a case the initial vector should be changed.

### 6.6.4 *Rate of convergence*

It is not difficult to prove that the sequence of vectors $z^{(k)}$ converges *linearly* to the eigenvector $x_{(1)}$. For we know that $z^{(k)} \simeq x_{(1)}$, and hence by applying the appropriate normalisation to equation (6.25) we deduce that

$$z^{(k)} = x_{(1)} + \frac{c_2(\lambda_2)^k}{c_1(\lambda_1)^k}x_{(2)} + \frac{c_3(\lambda_3)^k}{c_1(\lambda_1)^k}x_{(3)} + \cdots$$

and

$$z^{(k-1)} = x_{(1)} + \frac{c_2(\lambda_2)^{k-1}}{c_1(\lambda_1)^{k-1}}x_{(2)} + \frac{c_3(\lambda_3)^{k-1}}{c_1(\lambda_1)^{k-1}}x_{(3)} + \cdots$$

Since $|\lambda_2/\lambda_1| > |\lambda_3/\lambda_1| > \cdots$, it follows that, for sufficiently large $k$,

$$(x_{(1)} - z^{(k)}) \simeq \frac{\lambda_2}{\lambda_1}(x_{(1)} - z^{(k-1)}) \tag{6.27}$$

Hence the truncation error $x_{(1)} - z^{(k)}$ at iteration $(k)$ is a constant $\lambda_2/\lambda_1$ times the truncation error at iteration $(k-1)$, and the convergence is linear.

Clearly (Section 4.4.4 of Reference 1) the number of correct decimals gained at each iteration is

$$\log_{10}|\lambda_1/\lambda_2|$$

For example, for the matrix (6.23) we have obtained $\lambda_1 = 100$, and we also happen to know that $\lambda_2 = 1$. Hence the method should in theory gain 2 decimals per iteration, and this is exactly confirmed in the results, where $z^{(0)}$, $z^{(1)}$, $z^{(2)}$, $z^{(3)}$ are correct to 0, 2, 4, 6 decimals, respectively.

Note that convergence is very slow if $|\lambda_2|$ is close to $|\lambda_1|$. For example, if for a given problem it happened that $\lambda_1 = 1$ and $\lambda_2 = .9$, then the method would gain only 0.045 decimal places per iteration and no less than 22 iterations would be required for every correct decimal place!

## 6.7 Particular eigenvalues of a matrix

The smallest eigenvalue in magnitude, $\lambda_n$, of $A$ can be determined by simply applying the power method to the inverse matrix $A^{-1}$. For if $\lambda$, $x$ are an eigenvalue and eigenvector of $A$ then

$$Ax = \lambda x \Rightarrow A^{-1}Ax = \lambda A^{-1}x \Rightarrow A^{-1}x = (\lambda^{-1})x$$

and hence $\lambda^{-1}$, $x$ are an eigenvalue and eigenvector of $A^{-1}$. Clearly the largest value of $\lambda^{-1}$ corresponds to the smallest value of $\lambda$ and so the algorithm is as follows.

*Algorithm 6.6  Power method for the smallest eigenvalue*

(i) Determine $B = A^{-1}$ using Algorithm 5.4B.
(ii) Determine the largest eigenvalue $\mu$ and corresponding eigenvector $x$ of the matrix $B$ by Algorithm 6.5.
Then $\lambda = \mu^{-1}$ is the smallest eigenvalue of $A$, and $x$ is the corresponding eigenvector.

### 6.7.1  *Example of smallest eigenvalue*

Consider the matrix and starting vector

$$A = \begin{bmatrix} 5 & 4 & 0 \\ 6 & 5 & 0 \\ .01 & 0 & 1 \end{bmatrix}, \quad y^{(0)} = z^{(0)} = \begin{bmatrix} 1 \\ 1 \\ 1 \end{bmatrix} \quad (6.28)$$

Then by Algorithm 5.4B, or by calculating the matrix of cofactors of $A$, we deduce that

$$\mathbf{B} = \mathbf{A}^{-1} = \begin{bmatrix} 5 & -4 & 0 \\ -6 & 5 & 0 \\ -.05 & .04 & 1 \end{bmatrix}.$$

Then Algorithm 6.6 proceeds as follows:

$$\mathbf{y}^{(1)} = \mathbf{B}\mathbf{z}^{(0)} = \begin{bmatrix} 1 \\ -1 \\ .99 \end{bmatrix}, \alpha_1 = -1, \mathbf{z}^{(1)} = \begin{bmatrix} -1 \\ 1 \\ -.99 \end{bmatrix}$$

$$\mathbf{y}^{(2)} = \mathbf{B}\mathbf{z}^{(1)} = \begin{bmatrix} -9 \\ 11 \\ -.90 \end{bmatrix}, \alpha_2 = 11, \mathbf{z}^{(2)} = \begin{bmatrix} -.8182 \\ 1 \\ -.0818 \end{bmatrix}$$

$$\mathbf{y}^{(3)} = \mathbf{B}\mathbf{z}^{(2)} = \begin{bmatrix} -8.0910 \\ 9.9092 \\ -.0009 \end{bmatrix}, \alpha_3 = 9.9092, \mathbf{z}^{(3)} = \begin{bmatrix} -.8165 \\ 1 \\ -.0001 \end{bmatrix}$$

$$\mathbf{y}^{(4)} = \mathbf{B}\mathbf{z}^{(3)} = \begin{bmatrix} -8.0825 \\ 9.8990 \\ .0807 \end{bmatrix}, \alpha_4 = 9.8990, \mathbf{z}^{(4)} = \begin{bmatrix} -.8165 \\ 1 \\ .0082 \end{bmatrix}$$

$$\mathbf{y}^{(5)} = \mathbf{B}\mathbf{z}^{(4)} = \begin{bmatrix} -8.0825 \\ 9.8990 \\ .0890 \end{bmatrix}, \alpha_5 = 9.8990, \mathbf{z}^{(5)} = \begin{bmatrix} -.8165 \\ 1 \\ .0090 \end{bmatrix}$$

$$\mathbf{y}^{(6)} = \mathbf{B}\mathbf{z}^{(5)} = \begin{bmatrix} -8.0825 \\ 9.8990 \\ .0898 \end{bmatrix}, \alpha_6 = 9.8990, \mathbf{z}^{(6)} = \begin{bmatrix} -.8165 \\ 1 \\ .0091 \end{bmatrix}$$

Clearly $\mu = \lambda^{-1} = 9.8990$, and hence $\lambda = .1010$. Thus $\lambda = .1010$ and $\mathbf{x} = (-.8165 \ 1 \ .0091)^{\mathsf{T}}$ are the smallest eigenvalue and corresponding eigenvector to four decimal places.

Note that in this example the last component of $\mathbf{x}$ is not obtained correctly until the sixth iteration, although the first component of $\mathbf{x}$ and the eigenvalue $\lambda$ ($\simeq \alpha_k$) are correct by the third and fourth iteration. This illustrates that Algorithm 6.5 must not be terminated until *all* components of the vector $\mathbf{z}^{(k)}$ have converged.

A program for Algorithm 6.6 will not be given, since the following Algorithm 6.6A includes it as a special case (for $c = 0$).

### 6.7.2 *The eigenvalue nearest to a given value*

How can we determine eigenvalues other than the smallest and largest? If we have an idea of the approximate location of a particular eigenvalue, then we may modify the power method to exploit this. Suppose that $\lambda$ is the eigenvalue nearest to the value $c$, and that $\mathbf{x}$ is the corresponding eigenvector. Then

$$(\mathbf{A} - c\mathbf{I})\mathbf{x} = \mathbf{A}\mathbf{x} - c\mathbf{x} = \lambda\mathbf{x} - c\mathbf{x} = (\lambda - c)\mathbf{x}$$

and so $\lambda - c$ is the smallest eigenvalue of $\mathbf{A} - c\mathbf{I}$. It follows that $\mu = (\lambda - c)^{-1}$ is the largest eigenvalue of $(\mathbf{A} - c\mathbf{I})^{-1}$. We deduce the following algorithm, which includes Algorithm 6.6 as a special case for $c = 0$.

*Algorithm 6.6A Eigenvalue of $\mathbf{A}$ nearest to c*

(i) Determine $\mathbf{B} = (\mathbf{A} - c\mathbf{I})^{-1}$ using Algorithm 5.4B.
(ii) Determine the largest eigenvalue $\mu$ and corresponding eigenvector $\mathbf{x}$ of $\mathbf{B}$ by Algorithm 6.5.
   Then $\lambda = \mu^{-1} + c$ is the eigenvalue of $\mathbf{A}$ nearest to $c$ and $\mathbf{x}$ is the corresponding eigenvector.

*Example*

Consider again Example (6.28), and suppose that we require the eigenvalue of $\mathbf{A}$ nearest to $c = .9$. (This is obviously $\lambda_2 = 1$, corresponding to $\mathbf{x}_{(2)} = (0\ 0\ 1)^T$, and we use this merely as an illustration.) Then

$$(\mathbf{A} - c\mathbf{I}) = \begin{bmatrix} 4.1 & 4 & 0 \\ 6 & 4.1 & 0 \\ .01 & 0 & .1 \end{bmatrix},$$

and by Algorithm 5.4B,

$$\mathbf{B} = (\mathbf{A} - c\mathbf{I})^{-1} = \begin{bmatrix} -.5702 & .5563 & 0 \\ .8345 & -.5702 & 0 \\ .0570 & -.0556 & 10 \end{bmatrix}.$$

Now the power method for $\mathbf{B}$ proceeds as follows from $\mathbf{x}^{(0)} = (1\ 1\ 1)^T$:

$$\mathbf{y}^{(1)} = \mathbf{B}\mathbf{z}^{(0)} = \begin{bmatrix} -.0139 \\ .2643 \\ 10.0014 \end{bmatrix}, \quad \alpha_1 = 10.0014, \quad \mathbf{z}^{(1)} = \begin{bmatrix} -.0014 \\ .0264 \\ 1 \end{bmatrix}$$

$$\mathbf{y}^{(2)} = \mathbf{B}\mathbf{z}^{(1)} = \begin{bmatrix} .0155 \\ -.0163 \\ 9.9984 \end{bmatrix}, \; \alpha_2 = 9.9984, \; \mathbf{z}^{(2)} = \begin{bmatrix} .0016 \\ -.0016 \\ 1 \end{bmatrix}$$

$$\mathbf{y}^{(3)} = \mathbf{B}\mathbf{z}^{(2)} = \begin{bmatrix} -.0018 \\ .0022 \\ 10.0002 \end{bmatrix}, \; \alpha_3 = 10.0002, \; \mathbf{z}^{(3)} = \begin{bmatrix} -.0002 \\ .0002 \\ 1 \end{bmatrix}$$

$$\mathbf{y}^{(4)} = \mathbf{B}\mathbf{z}^{(3)} = \begin{bmatrix} .0002 \\ -.0003 \\ 10.0000 \end{bmatrix}, \; \alpha_4 = 10.0000, \; \mathbf{z}^{(4)} = \begin{bmatrix} .0000 \\ -.0000 \\ 1 \end{bmatrix}$$

Clearly    $\lambda = \mu^{-1} + c = (10.0000)^{-1} + 0.9 = 1.0000$

and    $\mathbf{x} = (0.0000 \; 0.0000 \; 1)^{\mathrm{T}}$.

Again we have linear convergence. In fact, the eigenvalues of $\mathbf{B}$ are:

$$\mu_1, \mu_2, \mu_2 = (1 - .9)^{-1}, (.101 - .9)^{-1}, (9.899 - .9)^{-1}$$
$$= \quad 10 \quad , \quad -1.252 \quad , \quad 0.111$$

Clearly the error must reduce at the constant rate

$$\mu_2/\mu_1 = \frac{-1.252}{10} = -.1252$$

and the algorithm must give

$$-\log_{10} |\mu_2/\mu_1| = .90 \text{ correct decimal places per iteration.}$$

This rate of convergence is confirmed in the numerical results above.
The BASIC program for Algorithm 6.6A now follows.

## Program 6.6A ENEAR: Power method for nearest eigenvalue

```
LIST
ENEAR

10    REM- ENEAR: FINDS EIGENVALUE NEAREST TO C AND CORRESP EIGENVECTOR
20    REM- OF MATRIX A.    USES POWER METHOD WITH NORMALISED VECTORS.
25    REM- ADDS C TO RECIPROCAL OF LARGEST EIGENVALUE OF INV(A-CI)
30    REM- ASSUMES LARGEST EIGENVALUE IN MAGNITUDE IS UNIQUE,SIMPLE,REAL.
40    REM- VECTORS: Y ITERATED, Z NORMALISED, Z1 PREVIOUS VALUE.
50    DIM Y(20),Z(20),Z1(20),A(20,20)
60    PRINT "DIMENSION OF A ";
70    INPUT N
80    PRINT "INPUT MATRIX A BY ROWS"
90    MAT INPUT A(N,N)
91    PRINT "NEAR WHAT VALUE DO YOU SEEK AN EIGENVALUE";
92    INPUT C
93    FOR I=1 TO N
94    A(I,I)=A(I,I)-C
95    NEXT I
96    MAT A=INV(A)
100   FOR I=1 TO N
110   Z1(I)=1
120   NEXT I
130   PRINT "ABSOLUTE ACCURACY REQUIRED";
140   INPUT E
```

```
150 PRINT "MAX NO OF ITERATIONS ALLOWED";
160 INPUT M
170 PRINT "DO YOU WANT TO CHOOSE YOUR OWN INITIAL VECTOR ?"
180 PRINT "INPUT 0 FOR NO , 1 FOR YES"
190 INPUT Q
200 IF Q=0 THEN 230
210 PRINT "INPUT INITIAL VECTOR:"
220 MAT INPUT Z1(N)
230 K=1
240 REM- ITERATION ON K BEGINS
250 REM- FORM Y=A*Z1, L=LARGEST ELEMENT, Z=Y/L.
260 D=0
270 L=0
280 FOR I=1 TO N
290 Y(I)=0
300 FOR J=1 TO N
310 Y(I)=A(I,J)*Z1(J)+Y(I)
320 NEXT J
330 F=ABS(Y(I))
340 IF D>F THEN 370
350 D=F
360 L=Y(I)
370 NEXT I
375 REM- L1 = APPROX EIGENVAL = C + 1/L
380 L1=C+1/L
390 PRINT "ITERATION";K
400 PRINT "EIGENVALUE:";L1
410 PRINT "EIGENVECTOR :"
420 FOR I=1 TO N
430 Z(I)=Y(I)/L
440 PRINT Z(I);
450 NEXT I
460 PRINT
470 REM- STOP IF MAX ABS (Z(I)-Z1(I))<=E  OR  K>=M
480 D=0
490 FOR I=1 TO N
500 F=ABS(Z(I)-Z1(I))
510 IF D>F THEN 530
520 D=F
530 Z1(I)=Z(I)
540 NEXT I
550 IF D<=E THEN 590
560 IF K>=M THEN 610
570 K=K+1
580 GO TO 260
590 PRINT "ABSOLUTE ACCURACY ACHIEVED"
600 GO TO 620
610 PRINT "MAX NO OF ITERATIONS REACHED"
620 END

Ready
```

## Sample run 1

```
RUN
ENEAR

DIMENSION OF A ? 3
INPUT MATRIX A BY ROWS
? 2.5,2.2,0.9,  1.1,2.5,1.1,  -1.1,2.5,2.4
NEAR WHAT VALUE DO YOU SEEK AN EIGENVALUE? 5
ABSOLUTE ACCURACY REQUIRED? .001
MAX NO OF ITERATIONS ALLOWED? 10
DO YOU WANT TO CHOOSE YOUR OWN INITIAL VECTOR ?
INPUT 0 FOR NO , 1 FOR YES
? 0
ITERATION 1
EIGENVALUE: 5.25161
EIGENVECTOR :
 .910555  .74  1
ITERATION 2
EIGENVALUE: 5.29691
EIGENVECTOR :
 .916646  .755439  1
ITERATION 3
EIGENVALUE: 5.29369
```

```
EIGENVECTOR :
 .916158  .754365  1
ITERATION 4
EIGENVALUE: 5.29392
EIGENVECTOR :
 .916197  .754439  1
ABSOLUTE ACCURACY ACHIEVED
Ready
```

## Sample run 2

```
RUN
ENEAR

DIMENSION OF A ? 3
INPUT MATRIX A BY ROWS
? 2.5,2.2,0.9,  1.1,2.5,1.1,  1.1,2.5,2.4
NEAR WHAT VALUE DO YOU SEEK AN EIGENVALUE? 1.5
ABSOLUTE ACCURACY REQUIRED? .001
MAX NO OF ITERATIONS ALLOWED? 10
DO YOU WANT TO CHOOSE YOUR OWN INITIAL VECTOR ?
INPUT 0 FOR NO , 1 FOR YES
? 0
ITERATION 1
EIGENVALUE: 2.29333
EIGENVECTOR :
-.400001  .133333  1
ITERATION 2
EIGENVALUE: 1.55183
EIGENVECTOR :
 1 -.914634E-01 -.910569
ITERATION 3
EIGENVALUE: 1.46348
EIGENVECTOR :
 1 -.100773 -.905352
ITERATION 4
EIGENVALUE: 1.46342
EIGENVECTOR :
 1 -.101022 -.904811
ABSOLUTE ACCURACY ACHIEVED
Ready
```

## Sample run 3

```
RUN
ENEAR

DIMENSION OF A ? 3
INPUT MATRIX A BY ROWS
? 2.5,2.2,0.9,  1.1,2.5,1.1,  1.1,2.5,2.4
NEAR WHAT VALUE DO YOU SEEK AN EIGENVALUE? .5
ABSOLUTE ACCURACY REQUIRED? .001
MAX NO OF ITERATIONS ALLOWED? 10
DO YOU WANT TO CHOOSE YOUR OWN INITIAL VECTOR ?
INPUT 0 FOR NO , 1 FOR YES
? 0
ITERATION 1
EIGENVALUE: 1.41528
EIGENVECTOR :
-.361111  1 -.625
ITERATION 2
EIGENVALUE: .669669
EIGENVECTOR :
-.694296  1 -.96964
ITERATION 3
EIGENVALUE: .64344
EIGENVECTOR :
-.709624  1 -.978157
ITERATION 4
EIGENVALUE: .642701
EIGENVECTOR :
-.710597  1 -.977857
ABSOLUTE ACCURACY ACHIEVED
Ready
```

*Program notes*

(1) This program may be formed by making the following amendments to Program 6.5.

    91 PRINT "NEAR WHAT VALUE DO YOU WANT AN
       EIGENVALUE";
    92 INPUT C
    93 FOR I = 1 TO N
    94 A(I, I) = A(I, I) − C
    95 NEXT I
    96 MAT A = INV(A)
    380 L1 = C + 1/L

Here instructions 91 to 96 change matrix **A** to matrix $\mathbf{A} - c\mathbf{I}$ and then invert it. The required eigenvalue is determined in instruction 380 as $c + \lambda^{-1}$, where $\lambda$ is the largest eigenvalue of $(\mathbf{A} - c\mathbf{I})^{-1}$. REM statement 10 is changed appropriately, and new REM statements 25 and 375 are included.

(2) For brevity, the built-in BASIC matrix routine is used to calculate $(\mathbf{A} - c\mathbf{I})^{-1}$ in instruction 96. Strictly speaking a more efficient 'subroutine' such as Algorithm 5.4B should be used here. However, this would make the program very long, since BASIC lacks the facility to call on a separate program (such as a program based on Algorithm 5.4B). The GOSUB facility is not really suitable, since it is only designed to cope with rather modest subprograms.

The FORTRAN language is in contrast very well able to introduce such subroutines, by way of the CALL statement. Once more we have encountered an unfortunate limitation in standard BASIC.

(3) In the three Sample runs, the program has found all the 3 eigenvalues and corresponding eigenvectors of a 3 by 3 matrix. This has been done by choosing 3 values of $c$, each of which is nearest to a different eigenvalue.

(4) It would be worthwhile and straightforward to modify the program so that any number of values of $c$ could be used to look for eigenvalues, without the necessity to read in a new **A**. This is left as an exercise to the reader (Problem 11). Note that A(I, J) would have to be copied into some variable A1(I, J), say, for all I, J, since the original values are destroyed in instructions 94 and 96.

## 6.8 Methods for all eigenvalues of a matrix

Let us suppose that all eigenvalues of **A** are real and of distinct magnitudes. Then an obvious strategy for obtaining all eigenvalues, and not restricting ourselves to largest, smallest, or nearest eigenvalues, is as follows. Having determined $\lambda_1$, the largest eigenvalue, and $\mathbf{x}_{(1)}$ the corresponding eigenvector, form a new matrix $\mathbf{A}^{(1)}$ of order $n - 1$ whose eigenvalues are precisely those of $\mathbf{A}$ but with $\lambda_1$ removed. Then determine the largest eigenvalue of $\mathbf{A}^{(1)}$, which is the second largest eigenvalue of $\mathbf{A}$, by the power method for $\mathbf{A}^{(1)}$, and so on. Such a process of successively reducing the size of $\mathbf{A}$ is called 'deflation'.

### 6.8.1 *Deflation of a matrix*

A deflation method is described in detail by C.E. Froberg (Reference 4). The matrix $\mathbf{A}^{(1)}$ is defined from $\mathbf{A}$ as

$$\mathbf{A}^{(1)} = \mathbf{A} - \mathbf{x}_{(1)}\mathbf{a}^{\mathrm{T}}$$

where $\mathbf{a}^{\mathrm{T}}$ is the first row of $\mathbf{A}$, and where $\mathbf{x}_{(1)}$ is normalised so that its first component is unity. The matrix $\mathbf{A}^{(1)}$ has largest eigenvalue $\lambda_2$ (and eigenvector $\mathbf{x}_{(1)} - \mathbf{x}_{(2)}$) and so the eigenvector $\mathbf{x}_{(2)}$ of $\mathbf{A}$ may be obtained by Algorithm 6.6A. Clearly the process can be repeated until all eigenvalues and eigenvectors of $\mathbf{A}$ have been determined.

This process will not, however, be discussed in any further detail. Unfortunately it is an *unstable* process, and so the calculated values of the largest eigenvalues of $\mathbf{A}$, $\mathbf{A}^{(1)}$, $\mathbf{A}^{(2)}$, ... become successively further and further away from the true values as the order of the matrix decreases. It is therefore wise to regard the deflation method primarily as one for finding crude approximations to eigenvalues and eigenvectors, which may be subsequently improved by other methods.

### 6.8.2 *Solution of the characteristic equation*

In principal the eigenvalues $\lambda$ of $\mathbf{A}$ can be determined by solving the 'characteristic equation' (compare (6.19))

$$f(\lambda) = |\mathbf{A} - \lambda\mathbf{I}| = 0 \qquad (6.29)$$

by using one or more of the algorithms discussed in Chapter 4 of Reference 1. Equation (6.29) can be treated as a nonlinear algebraic equation or more specifically as a polynomial equation. If the order $n$ is large, then it is not advisable to work out the coefficients of the polynomial $f(\lambda)$, but rather to evaluate $f(\lambda)$ by calculating the determinant of a matrix $\mathbf{A} - \lambda\mathbf{I}$ using the methods of Chapter 5. Nevertheless this is quite an expensive process, since about $\frac{1}{3}n^3$ multiplications are required for each evaluation of $f(\lambda)$.

Having solved (6.29) for a particular $\lambda$, the problem remains to determine the corresponding eigenvector $x$ from the eigenvalue problem equation (compare (6.20))

$$(A - \lambda I)x = 0 \qquad (6.30)$$

To solve such a homogeneous system, it is necessary to discard one equation, say the first, and fix $x_1 = 1$ (since any one non-zero component of $x$ may be chosen arbitrarily). There then remains a non-homogeneous system of $n - 1$ equations in $x_2, \ldots, x_n$ which may be solved by Gauss elimination. The formal algorithm is as follows:

*Algorithm 6.7   The characteristic equation method for eigenvalues and eigenvectors*

(i) Locate $n$ distinct approximate eigenvalues $\lambda^*$ of $A$ by applying the bisection method (Algorithm 4.1 of Reference 1) to solve the characteristic equation

$$f(\lambda) = |A - \lambda I| = 0$$

and calculating the determinant by Gauss elimination (Algorithm 5.4A).
For each approximate $\lambda^*$:

(ii) Starting from $\lambda^*$, determine an accurate eigenvalue $\lambda$ by using the secant method (Algorithm 4.3 of Reference 1) to solve $f(\lambda) = 0$ (or by using Algorithm 6.6A to find the nearest $\lambda$ to $\lambda^*$).

(iii) Determine the corresponding eigenvector by setting $x_1 = 1$ and solving by Gauss elimination (Algorithm 5.4A) the system

$$B^*x^* = b$$

where $B^*$ is the matrix $A - \lambda I$ minus its first row and column,

$x^*$ is the vector $x$ minus its first component,
$b = (-a_{21} - a_{31} \ldots - a_{n1})^T$.

If this fails, fix a different component of $x$ as unity and delete a different equation in (6.30).

We shall not give a specific BASIC program for this algorithm, but will leave it as an interesting project for an interested reader. Although it is not a commonly used method, it is nevertheless sound. Our main reason for presenting it here is that it can be implemented by making use exclusively of programs already developed. It is also worth pointing out that this method is readily extendable to the 'generalised eigenvalue problem'

$$(A + \lambda C)x = 0$$

where **A** and **C** are given square matrices, by solving for $\lambda$ the characteristic equation

$$f(\lambda) = |\mathbf{A} + \lambda\mathbf{C}| = 0$$

### 6.8.3 *The orthodox approach*

There are a number of more generally approved methods for obtaining *all* eigenvalues including Givens, Householder, Lanczos, and QR methods. These are beyond the scope of this book, but are discussed in References 3 and 4. They are all based on the idea of transforming the matrix **A** into a special form (such as a tridiagonal or 'Hessenberg' matrix) and then determining the eigenvalues by a special method.

## 6.9 References

1. Mason, J.C., *BASIC Numerical Mathematics*, Butterworths (1983).
2. Smith, G.D., *Numerical Solution of Partial Differential Equations*, Oxford University Press (1965).
3. Fox, L., *Numerical Linear Algebra*, Oxford University Press (1964).
4. Froberg, C.E., *Introduction to Numerical Analysis*, 2nd Edn., Addison-Wesley, London (1969).

## PROBLEMS

**(6.1)** Starting in each case from the initial vector $(1 - 1\ 1 - 1\ 1)^{\mathrm{T}}$, obtain the solution $(x_1\ x_2\ x_3\ x_4\ x_5)^{\mathrm{T}}$ of the system

$$\begin{bmatrix} -4 & 0 & 1 & 0 & 1 \\ 1 & -4 & 0 & 1 & 0 \\ 0 & 1 & -4 & 0 & 1 \\ 1 & 0 & 1 & -4 & 0 \\ 0 & 1 & 0 & 1 & -4 \end{bmatrix} \begin{bmatrix} x_1 \\ x_2 \\ x_3 \\ x_4 \\ x_5 \end{bmatrix} = \begin{bmatrix} -1 \\ 1 \\ -1 \\ 1 \\ -1 \end{bmatrix}$$

correct to 3 decimal places using
(i) the Jacobi iteration, (ii) the Gauss-Seidel iteration.
How many correct decimal places per iteration does the theory predict in case (i)?

**(6.2)** Consider the system $\mathbf{Ax} = \mathbf{b}$ given by

$$\begin{bmatrix} 2 & 1 & 1 \\ 1 & 2 & 1 \\ 1 & 1 & 2 \end{bmatrix} \begin{bmatrix} x_1 \\ x_2 \\ x_3 \end{bmatrix} = \begin{bmatrix} 4 \\ 4 \\ 4 \end{bmatrix}$$

(i) Does the theory tell us whether or not Jacobi's method converges to the solution $(1\ 1\ 1)^{\mathrm{T}}$, starting from an arbitrary initial vector?

Try Program 6.1 starting from $(.9 \ .9 \ .9)^T$. What happens?

(ii) Show that, for any $\mathbf{x}$,

$$\mathbf{x}^T \mathbf{A} \mathbf{x} = \sum_{i=1}^{3} \sum_{j=1}^{3} a_{ij} x_i x_j = x_1^2 + x_2^2 + x_3^2 + (x_1 + x_2 + x_3)^2$$

Deduce from the theory that the Gauss–Seidel method will converge in this case.

Try Program 6.3 starting from $(.9 \ .9 \ .9)^T$.

**(6.3)** Consider the system

$$\begin{bmatrix} 2 & 1 & 3 \\ 1 & 2 & 3 \\ 4 & 4 & 11 \end{bmatrix} \begin{bmatrix} x_1 \\ x_2 \\ x_3 \end{bmatrix} = \begin{bmatrix} 6 \\ 6 \\ 19 \end{bmatrix}$$

Does the theory tell us anything about the convergence of the Gauss–Seidel iteration in this case? Explain.

Test Programs 6.1 (Jacobi), 6.3 (Gauss–Seidel), and 6.4 (S.O.R.), and find an 'optimal' value of the relaxation factor $\omega$ to give the least number of iterations.

**(6.4)** Show that the recurrence relation

$$x_{i-1} - 20x_i + 3x_{i+1} = 0 \quad (i = 1, 2, 3, 4)$$

subject to $x_0 = 1$, $x_5 = 1$, leads to a tridiagonal system of equations for $x_1, x_2, x_3, x_4$. Use Program 6.2 to solve the system.

**(6.5)** Convert Program 6.2 into another program (GSTRI) which solves a tridiagonal system by the Gauss–Seidel method.

Test this new program on the system used in the sample run of Program 6.2.

**(6.6)** Write a new program which extends Program 6.2 so that it solves a 'multidiagonal' system of equations, in which the matrix has a specified number $2p + 1$ of diagonals of constant entries ($p$ above and $p$ below the main diagonal).

Test the program on Problem 4 and also on the 7 equations:

$$x_{i-2} - x_{i-1} + 16x_i - x_{i+1} + x_{i+2} = 0 \quad (i = 2, \ldots, 8)$$

where $x_0 = x_1 = x_9 = x_{10} = 1$. (*Determine* $x_2, \ldots, x_8$.)

**(6.7)** For a given matrix $\mathbf{A}$, define matrices $\mathbf{L}$, $\mathbf{D}$, $\mathbf{U}$:

$\mathbf{L} =$ matrix consisting of elements of $\mathbf{A}$ below its diagonal and zeros elsewhere

$\mathbf{D} =$ matrix consisting of diagonal elements of $\mathbf{A}$ and zeros elsewhere

$\mathbf{U} =$ matrix consisting of elements of $\mathbf{A}$ above its diagonal and zeros elsewhere i.e.

$$L = \begin{bmatrix} 0 & & \cdots & 0 \\ a_{21} & 0 & & \vdots \\ \vdots & & \ddots & \\ a_{n1} & \cdots & a_{n,n-1} & 0 \end{bmatrix}, \quad D = \begin{bmatrix} a_{11} & 0 & \cdots & 0 \\ 0 & a_{22} & & \vdots \\ \vdots & & \ddots & 0 \\ 0 & \cdots & \cdots & 0 & a_{nn} \end{bmatrix},$$

$$U = \begin{bmatrix} a_{11} & \cdots & a_{1n} \\ 0 & & \vdots \\ \vdots & \ddots & \\ 0 & \cdots & 0 & a_{nn} \end{bmatrix}$$

so that
$$L + D + U = A$$

Prove that the Jacobi, Gauss–Seidel, and S.O.R. methods for solving $Ax = b$ may be expressed in the respective forms:

Jacobi:　　　$x^{(k)} = D^{-1}b - D^{-1}(L + U)x^{(k-1)}$
Gauss–Seidel: $x^{(k)} = (L + D)^{-1}b - (L + D)^{-1}Ux^{(k-1)}$
S.O.R.:　　　$x^{(k)} = \omega(L + D)^{-1}b + [(1 - \omega)I - \omega(L + D)^{-1}U]x^{(k-1)}$

**(6.8)** Using similar analysis to that for Jacobi's method in Section 6.2.2, prove that the Gauss–Seidel method converges if the system matrix is strictly diagonally dominant.

Hint: Prove that $|e_r^{(k)}| \leqslant L.\max_i |e_i^{(k-1)}|$ $(r = 1, 2, \ldots, n)$ by induction on $r$ (proving the inequality first for $r = 1$). The analysis is thence identical to that of Section 6.2.2.

**(6.9)** Use the mathematical method of Section 6.5.1 to determine all eigenvalues and corresponding eigenvectors for the matrix

$$A = \begin{bmatrix} 4 & -1 & -1 \\ -1 & 2 & 1 \\ -1 & 1 & 2 \end{bmatrix}$$

[Answer: $5, (-2\ 1\ 1)^T$; $2, (1\ 1\ 1)^T$; $1, (0\ 1\ -1)^T$]

**(6.10)** Find the largest and smallest eigenvalues of the matrix $A$ of Problem 9, starting in each case from the vector $(0\ 1\ 0)^T$, and using Programs 6.5 and 6.6A.

**(6.11)** Write out a list of modifications necessary to Program 6.6A, so that control may be transferred at the end of the iteration back to instruction 91, and an eigenvalue may be found near a new value of $c$ (without inputting $A$ again). [Note that it is necessary to keep a copy of the original $A$, since this is destroyed in instructions 93 to 96.]

**(6.12)**
　(i) Suppose that $A$ is an $n$ by $n$ matrix. By considering the

characteristic equation $|\mathbf{A} - \lambda\mathbf{I}| = 0$ for eigenvalues $\lambda$, and determining the coefficients of $\lambda^n$ and $\lambda^{n-1}$, deduce that

$$\lambda_1 + \lambda_2 + \ldots + \lambda_n = a_{11} + a_{22} + \ldots + a_{nn}$$

(i.e. the sum of the eigenvalues is the sum of the diagonal terms).

(ii) The following foolproof algorithm may thus be used to find all eigenvalues (provided they are real and distinct) of a 3 by 3 matrix $\mathbf{A}$:

(a) Determine the largest and smallest eigenvalues $\lambda_1$ and $\lambda_3$ by Programs 6.5, 6.6A.

(b) Determine the intermediate eigenvalue as

$$\lambda_2 = a_{11} + a_{22} + a_{33} - \lambda_1 - \lambda_3$$

Use this algorithm to determine the 3 eigenvalues of the matrix used in the sample runs of Program 6.6A:

$$\mathbf{A} = \begin{pmatrix} 2.5 & 2.2 & 0.9 \\ 1.1 & 2.5 & 1.1 \\ 1.1 & 2.5 & 2.4 \end{pmatrix}$$

(6.13) Suppose that $\mathbf{A} = \begin{pmatrix} \sigma_{xx} & \sigma_{xy} & \sigma_{xz} \\ \sigma_{yx} & \sigma_{yy} & \sigma_{yz} \\ \sigma_{zx} & \sigma_{zy} & \sigma_{zz} \end{pmatrix}$

is the matrix of stresses at a point $P(x, y, z)$ in an elastic body (i.e. $\sigma_{xy}$ is the $y$-component of stress across a plane perpendicular to $0x$), and that $\mathbf{p} = (l\ m\ n)^{\mathrm{T}}$ is a unit vector (i.e. $l^2 + m^2 + n^2 = 1$) normal to some given plane through $P$.

(i) Show that the stress across this given plane will act along the normal if and only if $\mathbf{p}$ is an eigenvector of $\mathbf{A}$

i.e.                          $\mathbf{Ap} = \lambda\mathbf{p}$

Such a direction is called a 'principal stress direction' and $\lambda$ is called a 'principal stress'.

(ii) Find the principal stresses and principal stress directions at a point $P$ at which the matrix of stresses is

$$\begin{pmatrix} 3.9 & 0 & -1.1 \\ 0 & 1.1 & 0 \\ -1.1 & 0 & 4.1 \end{pmatrix}$$

[The principal stresses are close to 1, 3, 5.]

(6.14)
(i) If $\mathbf{x}_{(1)}, \mathbf{x}_{(2)}, \ldots, \mathbf{x}_{(n)}$ are eigenvectors of $\mathbf{A}$ ($n$ by $n$) and $\lambda_1$,

$\lambda_2, \ldots, \lambda_n$ are corresponding eigenvalues, show that the $n$ eigenvalue equations

$$\mathbf{Ax}_{(1)} = \lambda_1 \mathbf{x}_{(1)}, \ldots, \mathbf{Ax}_{(n)} = \lambda_n \mathbf{x}_{(n)}$$

may be written collectively in the matrix form

$$\mathbf{AK} = \mathbf{KD}$$

where $\mathbf{K} = (\mathbf{x}_{(1)} | \mathbf{x}_{(2)} | \ldots | \mathbf{x}_{(n)})$   (matrix of eigenvectors) and

$$\mathbf{D} = \begin{bmatrix} \lambda_1 & 0 & & 0 \\ 0 & \lambda_2 & & \vdots \\ & & & 0 \\ 0 & \ldots & 0 & \lambda_n \end{bmatrix} \quad \begin{array}{l} \text{(diagonal matrix} \\ \text{of eigenvalues)} \end{array}$$

(ii) Show that the matrix  $\mathbf{A} = \begin{bmatrix} 3 & 0 & -1 \\ 0 & 3 & 0 \\ -1 & 0 & 3 \end{bmatrix}$

has eigenvalues and eigenvectors:

$$3, (0\ 1\ 0)^{\mathrm{T}};\quad 2, \left(\frac{1}{\sqrt{2}}\ 0\ \frac{1}{\sqrt{2}}\right)^{\mathrm{T}};\quad 4, \left(\frac{1}{\sqrt{2}}\ 0\ -\frac{1}{\sqrt{2}}\right)^{\mathrm{T}}$$

and verify the relationship $\mathbf{AK} = \mathbf{KD}$ in this case.

(iii) Show that the matrix $\mathbf{K}$ of part (ii) has the property that $\mathbf{K}^{\mathrm{T}}\mathbf{K} = \mathbf{I}$. Such a matrix is called 'orthogonal'.

**(6.15)**

(i) Show that the system of *coupled* simultaneous differential equations

$$\begin{array}{llll} \ddot{y}_1 + 3y_1 & - y_3 = 0; & y_1(0) = \sqrt{2}, & \dot{y}_1(0) = 0 \\ \ddot{y}_2 & + 3y_2 & = 0; & y_2(0) = 2, & \dot{y}_2(0) = 0 \\ \ddot{y}_3 - y_1 & + 3y_3 = 0; & y_3(0) = 0, & \dot{y}_3(0) = 0 \end{array}$$

may be written collectively in the matrix form

$$\ddot{\mathbf{y}} + \mathbf{Ay} = \mathbf{0};\quad \mathbf{y}(0) = \mathbf{b} = (\sqrt{2}\ 2\ 0)^{\mathrm{T}},\quad \dot{\mathbf{y}}(0) = \mathbf{0}$$

where $\mathbf{y} = (y_1\ y_2\ y_3)^{\mathrm{T}}$, $\dot{\mathbf{y}} = (\dot{y}_1\ \dot{y}_2\ \dot{y}_3)^{\mathrm{T}}$, etc. and $\mathbf{A}$ is the matrix of Problem 14(ii).

(ii) Show that the change of variables $\mathbf{y} = \mathbf{Kz}$,

i.e.
$$\begin{cases} y_1 = \dfrac{1}{\sqrt{2}} z_2 + \dfrac{1}{\sqrt{2}} z_3 \\[2mm] y_2 = z_1 \\[2mm] y_3 = \dfrac{1}{\sqrt{2}} z_2 - \dfrac{1}{\sqrt{2}} z_3 \end{cases}$$

where $\mathbf{K}$ is the matrix of eigenvectors of $\mathbf{A}$ (as in Problem 14(ii)), takes the system into the *uncoupled* system of equations of simple harmonic motion

$$\ddot{z}_1 + \lambda_1 z_1 = 0; \quad z_1(0) = \beta_1, \; \dot{z}_1(0) = 0$$
$$\ddot{z}_2 + \lambda_2 z_2 = 0; \quad z_2(0) = \beta_2, \; \dot{z}_2(0) = 0$$
$$\ddot{z}_3 + \lambda_3 z_3 = 0; \quad z_3(0) = \beta_3, \; \dot{z}_3(0) = 0$$
i.e. $\ddot{\mathbf{z}} + \mathbf{Dz} = \mathbf{0}; \quad \mathbf{z}(0) = \boldsymbol{\beta}, \; \dot{\mathbf{z}}(0) = \mathbf{0}$

where $\lambda_1, \lambda_2, \lambda_3$ are the eigenvalues of $\mathbf{A}$ and $\boldsymbol{\beta} = (\beta_1 \; \beta_2 \; \beta_3)^{\mathrm{T}}$ is the solution of $\mathbf{K}\boldsymbol{\beta} = \mathbf{b}$.

(iii) Verify that $\boldsymbol{\beta} = (2 \; 1 \; 1)^{\mathrm{T}}$ and hence, by solving the above simple harmonic motion equations for $z_1, z_2, z_3$, deduce that

$$y_1 = \frac{1}{\sqrt{2}} (\cos \sqrt{2}t + \cos 2t)$$

$$y_2 = 2 \cos \sqrt{3}t$$

$$y_3 = \frac{1}{\sqrt{2}} (\cos \sqrt{2}t - \cos 2t)$$

This problem determines the three-dimensional vibration of a body which is released from rest in a displaced position.

**(6.16)** By using a similar method to that of Problem 15, and noting that the solution of

$$\ddot{z}_i + \lambda_i z_i = 0; \; z_i(0) = 0, \; \dot{z}_i(0) = a_i$$

is

$$z_i = \frac{a_i}{\sqrt{\lambda_i}} \sin \sqrt{\lambda_i}t$$

determine the solution $\mathbf{y}$ of the coupled system

$$\ddot{\mathbf{y}} + \mathbf{Ay} = \mathbf{0}; \quad \mathbf{y}(0) = \mathbf{0}, \; \dot{\mathbf{y}}(0) = \mathbf{a}$$

where (i) $\mathbf{A}$ is the matrix of Problem 9 and $\mathbf{a} = (-1 \; 4 \; 0)^{\mathrm{T}}$, (ii) $\mathbf{A}$ is the matrix of Problem 12(ii) and $\mathbf{a} = (1 \; 1 \; 1)^{\mathrm{T}}$.

Case (i) may easily be written down exactly. Case (ii) requires the use of Program 6.6A to determine $\lambda_1, \lambda_2, \lambda_3$ and Program 5.4A to solve $\mathbf{K}\boldsymbol{\alpha} = \mathbf{a}$ for $\boldsymbol{\alpha} = \dot{\mathbf{z}}(0)$.

# Chapter 7

# Matrix methods in approximation and data fitting

## ESSENTIAL THEORY

There are many significant areas of application of matrix methods, and especially of the methods of Chapters 5 and 6, and two particularly important ones are the numerical solution of differential equations and approximation and data fitting. We have chosen to consider the latter area, since it seems more appropriate for a brief discussion.

Emphasis is given to collocation and least squares methods for generalised linear approximation and in particular, for polynomial and spline approximation. In order to provide unity to the chapter, a brief introduction to the general subject area is given in Sections 7.1 to 7.3, before embarking on collocation methods in Section 7.4, least squares methods in Section 7.5, and spline function methods in Section 7.6.

## 7.1 Introduction

It is advantageous and indeed often necessary to be able to represent a given mathematical function by a simpler function whose values are approximately the same. For example, $e^x$ can be represented on the range $[0, 1]$ to an accuracy of 0.00060 by a suitable cubic in $x$ (see Sample run 2 of Program 7.1). The simpler function, called an *approximation* or *approximating function*, is effectively an inexpensive substitute for the original. Designers of computer software invariably use approximations in constructing built-in subroutines for standard functions like $e^x$, $\cos x$, $\log x$, etc, and engineers and scientists often exploit them to represent functions of their own.

Incidentally, an approximation is not the only way of representing a function. An alternative, which is quite commonly used today and was almost universally used in pre-computer days, is to set out values of the function in a *table*. The table should be detailed enough so that simple *interpolation* procedures can be used to calculate values between those tabulated (see Chapter 5 of Reference 1). However, tables need to be voluminous if simple straight line interpolation is to be used, and examples of such tables may be found among earlier publications of establishments such as the National Bureau of Standards in the USA and the National Physical Laboratory in the UK.

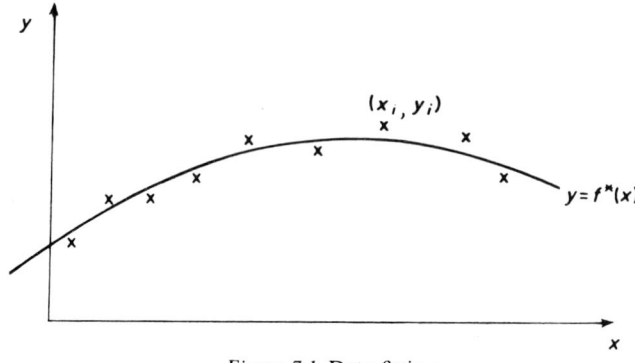

*Figure 7.1.* Data fitting.

With the advent of computers, approximations have tended to replace large tables for reasons of convenience and economy.

Having decided that approximations are needed, we must learn how to find them; and having found them, we must assess how good they are. These two tasks are our primary concern in this chapter.

A topic closely related to function approximation is that of *data fitting*, that is to say fitting a smooth curve to a given set of data (see Figure 7.1) If there are no appreciable experimental or other errors in the data, then the smooth curve is essentially an approximation to a function defined only at a *discrete set* of points (in contrast to a mathematical function like $e^x$ which is defined at all points of an *interval*). Even if experimental errors in the data are appreciable, simple approximations may still be used. Indeed it is common in such cases to assume some kind of *statistical distribution* of the data errors, such as a normal distribution, and to exploit this in the subsequent method of approximation. It is even possible to a certain extent to smooth out some of the data errors as an integral part of the method.

## 7.2  General setting for approximation

Both approximation and data fitting problems may be posed formally in the setting of the following common 'approximation problem':

Given (i) a function $f(x)$ defined either on an interval $[a, b]$ or on a discrete point set $x_1, x_2, \ldots, x_m$, (ii) a form of approximation

$f^*(x)$ involving $n$ undetermined parameters $c_1, c_2, \ldots, c_n$, (iii) a norm of approximation $\|f - f^*\|$ which measures the error in approximating $f$ by $f^*$ on $[a, b]$ or on $\{x_1, x_2, \ldots, x_m\}$, determine values of the parameters $c_1, \ldots, c_n$ in $f^*$ so that $\|f - f^*\|$ is 'acceptably small'.

Although the idea of a 'norm', which will be discussed shortly, is probably new to the reader, the specific choices which are most commonly made will certainly be familiar to him.

The phrase 'acceptably small' might tempt the reader to retort 'Unacceptably vague!' However, on reflection, he will probably agree that it is ultimately up to the user to decide what approximation error he is prepared to accept. Nevertheless, it is possible to be more precise by posing the following idealistic 'best approximation problem':

Given (i), (ii), (iii) (as in the 'approximation problem'), determine values of the parameters $c_1, c_2, \ldots, c_n$ in $f^*$ so that $\|f - f^*\|$ has its minimum value over all choices of parameters.

The resulting approximation $f^*$ is called a 'best approximation'.

Although the best approximation problem is sometimes too expensive and time-consuming to solve, it always provides something useful to aspire to.

### 7.2.1 Choice of form and norm

Before attempting to solve the approximation problem, an appropriate form and norm must be chosen. The choice of a good form can be difficult, and it tends to be as much an art as a science. Three commonly used forms are as follows:

(i) Polynomial: $f^*(x) = c_1 + c_2 x + \ldots + c_n x^{n-1}$

(ii) Rational function: $f^*(x) = \dfrac{c_1 + c_2 x + \ldots + c_p x^{p-1}}{1 + c_{p+1} x + \ldots + c_n x^{n-p}}$

This is a ratio of 2 polynomials.

(iii) Spline function: $f^*(x) = c_1 + c_2 x + c_p x^{p-1}$
$$+ c_{p+1}(x - z_1)_+^{p-1} +$$
$$\ldots + c_n(x - z_{n-p})_+^{p-1}$$
where $(x - a)_+^{p-1} = \begin{cases} (x - a)^{p-1} & \text{for } x \geqslant a \\ 0 & \text{for } x \leqslant a \end{cases}$

Here $z_1, \ldots, z_{n-p}$ are the *knots* of the spline, and the number $p - 1$ is its *degree*.

In this chapter we shall initially concentrate on the form (i), since it is the most fundamental, but we shall later discuss form (iii) which has great practical importance. It will also be advantageous to consider the generalised form

$$f^*(x) = c_1 g_1(x) + c_2 g_2(x) + \ldots + c_n g_n(x)$$

where $g_j(x)(j = 1, \ldots, n)$ are specified functions, since this includes (i) and (iii) as special cases, and permits many other useful forms to be covered.

If we had to choose one of the forms (i) to (iii), which would it be? The answer would depend on the nature of $f(x)$. Indeed in Figure 7.2 are illustrated three functions (a), (b), (c) for which the respective forms (i), (ii), (iii) are the most promising candidates. Broadly speaking, a polynomial is a naturally 'roly-poly' function, without sudden jumps or prolonged 'flat' behaviour. It would therefore make an obvious choice for (a) and an unlikely choice for (b) or (c). A rational function can be defined over an infinite range of $x$, and so it is the best candidate to fit the decaying function (b). Although it is quite a versatile form, it needs to be used with some caution, since there is always the possibility of it having a pole (i.e. a zero in its denominator). A spline function is very versatile and is really the only form which copes consistently well with a function like (c). This function is ideal to approximate in pieces, and a spline does just that, since its nature changes each time $x$ passes a knot.

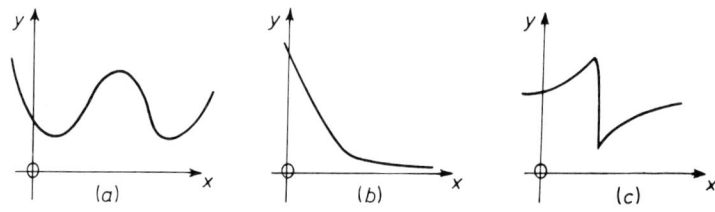

*Figure* 7.2. Typical functions.

The choice of a norm is usually more straightforward and we can make some broad statements which cover most common cases. If the function is known *exactly* then the most appropriate norm is generally the *maximum absolute error* or *Chebyshev norm* (written $\|.\|_\infty$):

$$\|f - f^*\|_\infty = \left. \begin{array}{l} \displaystyle\max_{a \leqslant x \leqslant b} |f(x) - f^*(x)| \text{ on an interval } [a, b] \\ \text{or } \displaystyle\max_{1 \leqslant i \leqslant m} |f(x_i) - f^*(x_i)| \text{ on a discrete set } \{x_i\} \end{array} \right\} \quad (7.1)$$

This norm has the advantage that it is a 'natural' measure of approximation which is not exceeded at any point, and the disadvantage that the best approximation can normally only be obtained by an iterative procedure (see Reference 2). If the data have errors which are not negligible, but which lie in a statistical distribution to the *normal distribution*, then the most appropriate norm is the *root mean square error* or *least squares norm* (written $\|.\|_2$):

$$\|f-f^*\|_2 = \left[\int_a^b \{f(x)-f^*(x)\}^2 \, dx\right]^{1/2} \quad \text{on an interval } [a, b]$$

$$\text{or} \left[\frac{1}{m}\sum_{i=1}^m \{f(x_i) \qquad -f^*(x_i)\}^2\right]^{1/2} \quad \text{on a discrete set } \{x_i\} \tag{7.2}$$

This choice of norm has two advantages: firstly it permits an easy determination of a best approximation, and secondly it is directly related to an 'unbiased' estimate of the variance of the recorded data from their true values.

Our discussion would not be complete without mention of a third norm, which is useful in special cases, namely the *mean error* or *least first power norm* (written $\|.\|_1$):

$$\|f-f^*\|_1 = \int_a^b |f(x)-f^*(x)| \, dx \quad \text{or} \quad \frac{1}{m}\sum_{i=1}^m |f(x_i)-f^*(x_i)| \tag{7.3}$$

This norm is particularly useful in the case of discrete data if there is a possibility that isolated 'wild' points may occur (at which the error is much larger than elsewhere), since such points tend to be given little or no emphasis. The best approximation problem is not easy for this norm; it involves a repetitive procedure, which is based on linear programming in the discrete case (see Reference 2).

## 7.3 Chebyshev polynomials and their properties

Let us first consider approximating a function $f(x)$ by a polynomial $f^*(x)$ in the Chebyshev norm $\|f-f^*\|_\infty$ (i.e. the maximum absolute error). Then we are inevitably led to a system of polynomials, again named after the famous Russian mathematician, Chebyshev.

The Chebyshev polynomial $T_n(x)$ of degree $n$ is defined by relating it to the cosine function of another variable $\theta$:

$$T_n(x) = \cos n\theta \quad \text{where} \quad x = \cos\theta. \tag{7.4}$$

More compactly, but less simply, we may write

$$T_n(x) = \cos(n\cos^{-1}x). \tag{7.5}$$

Now

$\cos 0.\theta = 1, \cos 1.\theta = \cos\theta, \cos 2\theta = 2\cos^2\theta - 1,$
$\cos 3\theta = 4\cos^3\theta - 3\cos\theta, \cos 4\theta = 8\cos^4\theta - 8\cos^2\theta + 1,$ etc.

Indeed it is left as an exercise (Problem 2) for the reader to deduce that, for any integer $n$, $\cos n\theta$ can be written as a polynomial of degree $n$ in $\cos\theta$ in which the leading coefficient (of $\cos^n\theta$) is $2^{n-1}$. It follows immediately that $T_n(x)$ is a polynomial of degree $n$ in $x$ with leading coefficient $2^{n-1}$. Specifically,

$$\left.\begin{array}{l} T_0(x) = 1, \; T_1(x) = x, \; T_2(x) = 2x^2 - 1, \; T_3(x) = 4x^3 - 3x \\ T_4(x) = 8x^4 - 8x^2 + 1, \text{ etc.} \end{array}\right\} \quad (7.6)$$

Many formulae are available in the literature in connection with the Chebyshev polynomials, including algorithms to determine the coefficients of the various powers of $x$ in $T_n(x)$, and the reader is referred to the text of Snyder (Reference 3) for details.

The Chebyshev polynomial $T_n(x)$ has been defined here for use on the range $[-1, 1]$ of $x$. However, we are frequently interested instead in some other range of $x$, say $[a, b]$. In that case we use instead of $T_n(x)$ the polynomial

$$T_n\left(\frac{2}{b-a}\left\{x - \frac{(a+b)}{2}\right\}\right) \quad (7.7)$$

i.e.     $T_n(z)$ where $z = \dfrac{2}{b-a}\left\{x - \dfrac{(a+b)}{2}\right\}$     $(7.8)$

since it is clear that $z = -1$, $1$ correspond to $x = a$, $b$. In the important case $a = 0$, $b = 1$, in which the range of approximation is $[0, 1]$, the polynomial (7.7) is denoted by $T_n^*(x)$ and takes the form

$$T_n^*(x) = T_n(2x - 1)$$

We refer to $T_n^*(x)$ as the *shifted Chebyshev polynomial* of degree $n$, and specific formulae are

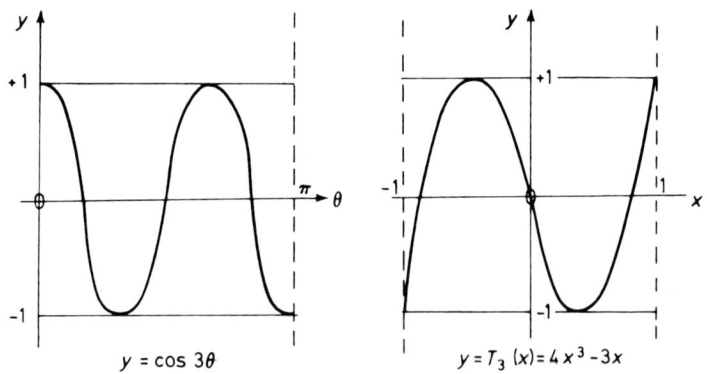

*Figure* 7.3. $\cos 3\theta$ and $T_3(x)$.

$$T_0^*(x) = 1, \; T_1^*(x) = 2x - 1, \; T_2^*(x) = 8x^2 - 8x + 1, \text{ etc}$$

For an idea of the (oscillatory) behaviour of the Chebyshev polynomials, the reader is referred to Figure 7.3 where $T_n(x)$ and $\cos n\theta$ are compared for $n = 3$.

### 7.3.1 Chebyshev polynomial zeros

It is clear from Figure 7.3 that $T_3(x)$ is zero at 3 distinct points $x$ in $[-1, 1]$, and indeed, for any $n$, $T_n(x)$ has exactly $n$ zeros in $[-1, 1]$. These zeros, which prove useful in Section 7.4 below, are easily determined from the relations

$$T_n(x) = \cos n\theta = 0 \text{ where } x = \cos \theta \; (-1 \leqslant x \leqslant 1, 0 \leqslant \theta \leqslant \pi).$$

Clearly $\cos n\theta$ is zero when the angle $\theta$ is an odd multiple of $90°$, and so

$$n\theta = (i - \tfrac{1}{2})\pi \quad (i = 1, 2, 3, \ldots)$$

To limit values of $\theta$ to the range $[0, \pi]$, we choose

$$\theta = (i - \tfrac{1}{2})\pi/n \quad (i = 1, 2, \ldots, n)$$

The $n$ distinct zeros of $T_n(x)$ are thus the corresponding $x$ values:

$$x = x_i = \cos\{(i - \tfrac{1}{2})\pi/n\} \quad (i = 1, 2, \ldots, n) \tag{7.9}$$

i.e. $\quad x = x_1, x_2, \ldots, x_n = \cos\dfrac{\pi}{2n}, \cos\dfrac{3\pi}{2n}, \ldots, \cos\dfrac{(2n-1)\pi}{2n}$

For example the 3 zeros of $T_3(x)$ are

$$x = x_1, x_2, x_3 = \cos\frac{\pi}{6}, \cos\frac{3\pi}{6}, \cos\frac{5\pi}{6} = \frac{\sqrt{3}}{2}, 0, \frac{-\sqrt{3}}{2}$$

We shall also require the Chebyshev zeros appropriate to a general interval $[a, b]$ of $x$. From (7.7), it is easy to deduce the general formula (consistent with (7.9) for $a = -1, b = 1$):

$$x = x_i = \frac{a + b}{2} + \frac{b - a}{2} \cos\{(i - \tfrac{1}{2})\pi/n\} \quad (i = 1, \ldots, n). \tag{7.10}$$

### 7.3.2 The alternating property, and the minimax property

We now give two key properties of $T_n(x)$; the first is geometrical and the second is analytical (and may be deduced from the first). Briefly they tell us that $T_n(x)$ oscillates up and down as often as possible between $+1$ and $-1$, and hence that it is smaller than any other polynomial of degree $n$ with the same leading coefficient. The second property is called the 'minimax' property, since it establishes a 'minimum maximum' for $T_n(x)$. The proofs are omitted, but are discussed in Problem 3.

*Property 1 (Alternating Property).* On the range $[-1, 1]$, $T_n(x)$ has a maximum magnitude of 1 which it attains with alternating signs at precisely $n + 1$ consecutive points

$$x = -1, \frac{\cos(n-1)\pi}{n}, \frac{\cos(n-2)\pi}{n}, \ldots, \frac{\cos 2\pi}{n}, \frac{\cos \pi}{n}, 1$$

i.e.         at $x = \dfrac{\cos(n-i)\pi}{n}$    $(i = 0, 1, \ldots, n)$

*Property 2 (Minimax Property).* On the range $[-1, 1]$, the polynomial $2^{1-n}T_n(x)$ has a smaller maximum magnitude, namely $2^{1-n}$, than any other monic polynomial of degree $n$. [A monic polynomial $\phi_n(x)$ of degree $n$ is a polynomial whose leading coefficient is unity, i.e. $\phi_n(x) = x^n + b_1 x^{n-1} + \ldots + b_n$ for some $b_1, \ldots, b_n$.]

### 7.3.3 Examples of best approximation in the Chebyshev norm

Property 2 enables us to give an example of a function $f(x)$ for which the best approximation $f^*(x)$ may be determined. Suppose that

$$f(x) = x^n, \text{ and } f^*(x) = c_1 + c_2 x + \ldots + c_n x^{n-1}$$

i.e. $f^*(x)$ is a polynomial of one degree lower than $f(x)$.
Then $f(x) - f^*(x)$ is a monic polynomial of degree $n$, and $\| f - f^* \| = \max\limits_{-1 \leqslant x \leqslant 1} |f(x) - f^*(x)|$. From Property 2, we immediately deduce that $f^*(x)$ is the best approximation if

$$f(x) - f^*(x) = 2^{1-n}T_n(x)$$

and hence   $f^*(x) = f(x) - 2^{1-n}T_n(x) = x^n - 2^{1-n}T_n(x)$

For example, the best approximation to $x^4$ on $[-1, 1]$ by a cubic polynomial is

$$f^*(x) = x^4 - 2^{-3}T_4(x) = x^4 - \tfrac{1}{8}(8x^4 - 8x^2 + 1) = x^2 - 0.125$$

which is in fact a quadratic. Since $T_n(x)$ has a maximum magnitude of unity (by Property 1), the maximum error in replacing $x^4$ by $x^2 - 0.125$ on $[-1, 1]$ is

$$\| f - f^* \| = \max\limits_{-1 \leqslant x \leqslant 1} |x^4 - (x^2 - 0.125)| = \max \{2^{-3}|T_4(x)|\} = 2^{-3}$$

$$= 0.125$$

Thus $x^4$ is represented to an absolute accuracy of 0.125 at all points of $[-1, 1]$ by the quadratic $x^2 - 0.125$, and this is illustrated in Figure 7.4. Note that the error has five extreme values of equal

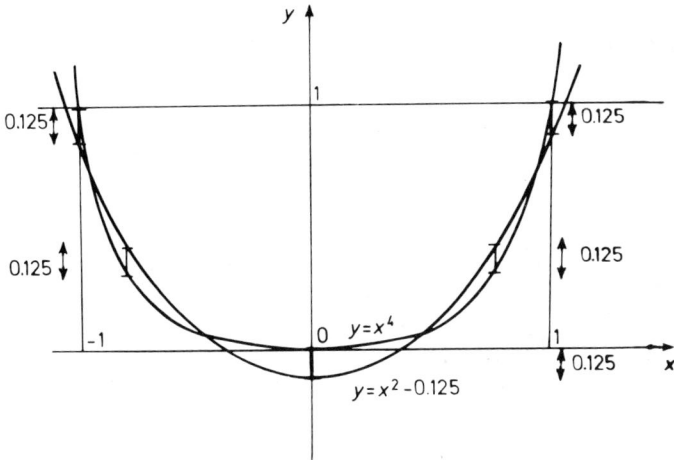

*Figure 7.4.* $f(x) = x^4$ approximated by $f^*(x) = x^2 - 0.125$.

magnitudes (0.125) and alternating signs. Such an 'equal oscillation' property of the error is in fact achieved by the best polynomial approximation to *any* continuous function $f(x)$.

In general, unfortunately, there is no simple method of forming a best approximation in the Chebyshev norm to a given function $f(x)$, and indeed all of the available algorithms are iterative ones (see Reference 2 for details). However, it is possible by direct methods to obtain approximations which are nearly as accurate as best approximations, and in particular the collocation method, which is a relatively simple procedure to implement, sometimes leads to such 'near-best' approximations.

## 7.4 Approximation by collocation

Collocation is the procedure of exactly matching an $n$-parameter approximation $f^*(x)$ to a given function $f(x)$ at $n$ specified points. For example the straight line $f^*(x) = c_1 + c_2 x$ collocates the function $f(x) = e^x$ at $x = 0, 1$ if and only if

$$c_1 = 1, c_2 = e^{-1}$$

The word 'interpolation' is commonly used in place of 'collocation'; but to avoid ambiguity, we shall only use 'interpolation' for its more fundamental meaning of 'inserting values between other values' (see Reference 1).

In the case of the polynomial form

$$f^*(x) = c_1 + c_2x + \ldots + c_nx^{n-1} \tag{7.11}$$

we collocate at $n$ distinct points $x = x_1, \ldots, x_n$. Then $c_1, \ldots, c_n$ are given by the *collocation equations* :

$$f^*(x) = f(x) \text{ at } x = x_i \quad (i = 1, \ldots, n) \tag{7.12}$$

i.e.    $c_1 + c_2x_i + c_3(x_i)^2 + \ldots + c_n(x_i)^{n-1} = f(x_i) \quad (i = 1, \ldots, n)$

$$\tag{7.13}$$

The equations (7.13) form a linear system of n equations in $n$ unknowns. It is easy to verify that the determinant of the left-hand side is non-zero (see Section 3.10), and hence that there is unique solution for $c_1, \ldots, c_n$. For example, for n = 3, (7.13) gives the three simultaneous equations:

$$c_1 + c_2x_1 + c_3x_1{}^2 = f(x_1)$$
$$c_1 + c_2x_2 + c_3x_2{}^2 = f(x_2)$$
$$c_1 + c_2x_3 + c_3x_3{}^2 = f(x_3)$$

and hence
$$\begin{bmatrix} 1 & x_1 & x_1{}^2 \\ 1 & x_2 & x_2{}^2 \\ 1 & x_3 & x_3{}^2 \end{bmatrix} \begin{bmatrix} c_1 \\ c_2 \\ c_3 \end{bmatrix} = \begin{bmatrix} f(x_1) \\ f(x_2) \\ f(x_3) \end{bmatrix}$$

Clearly these may be written in matrix form

$$\mathbf{Ac} = \mathbf{f}$$

where $\mathbf{A}$, $\mathbf{c}$, and $\mathbf{f}$ correspond to the arrays in the previous equation, and the solution $\mathbf{c}$ may be obtained by the methods of Chapter 5.

The same idea may be used for the generalised approximation:

$$f^*(x) = c_1g_1(x) + c_2g_2(x) + \ldots + c_ng_n(x) = \sum_{j=1}^{n} c_jg_j(x) \tag{7.14}$$

where $g_1(x), g_2(x), \ldots, g_n(x)$ are any set of $n$ specified 'basis' functions. In order for (7.14) to genuinely involve $n$ parameters, the functions $g_1(x), \ldots, g_n(x)$ must be chosen to be *linearly independent*, so that no one of them is a linear combination of the others. The polynomial form (7.11) is a special case of (7.14) corresponding to the choice

$$g_j(x) = x^{j-1} \quad (j = 1, \ldots, n)$$

Other possible choices might, for example, be $g_j(x) = \sin j\pi x$ or $g_j(x) = \exp[(j-1)x]$. For convenience in later BASIC programming, we define a function $g(x, j)$ of two variables by the relation

$$g(x, j) = g_j(x) \qquad (7.15)$$

In the case of a polynomial, $g(x, j) = x^{j-1}$. Clearly we can meaningfully allow $x$ and $j$ to take any real values, although in fact $j$ will only be given the integer values 1, 2, ..., $n$.

Substituting (7.14) into the collocation equations (7.12)

$$\sum_{j=1}^{n} c_j g(x_i, j) = f(x_i) \quad (i = 1, \ldots, n) \qquad (7.16)$$

In matrix form this may be rewritten as

$$\mathbf{Ac} = \mathbf{f} \qquad (7.17)$$

where

$$\mathbf{A} = \begin{bmatrix} g(x_1, 1) & g(x_1, 2) & \cdots & g(x_1, n) \\ g(x_2, 1) & g(x_2, 2) & \cdots & g(x_2, n) \\ \vdots & \vdots & & \vdots \\ g(x_n, 1) & g(x_n, 2) & \cdots & g(x_n, n) \end{bmatrix}, \; \mathbf{c} = \begin{bmatrix} c_1 \\ c_2 \\ \vdots \\ c_n \end{bmatrix}, \; \mathbf{f} = \begin{bmatrix} f(x_1) \\ f(x_2) \\ \vdots \\ f(x_n) \end{bmatrix}$$

$$(7.18)$$

More briefly, if the entry in row $i$ of $\mathbf{f}$ is $f_i$ and the entry in row $i$ and column $j$ of $\mathbf{A}$ is $A_{ij}$, then

$$A_{ij} = g(x_i, j), \; f_i = f(x_i) \quad (i = 1, \ldots, n; \quad j = 1, \ldots, n) \; (7.19)$$

Since $\mathbf{A}$ and $\mathbf{f}$ are known, the parameters $\mathbf{c}$ may be obtained by solving the linear algebraic system (7.17). For this purpose the Gauss elimination method and the corresponding programs of Chapter 5 may be used.

The following algorithm and program perform generalised collocation. In the absence of a suitable subprogram facility in BASIC, and for simplicity of programming, the BASIC matrix inversion routine is used to solve (7.17).

*Algorithm 7.1 Generalised approximation by collocation*
To determine an approximation $\Sigma c_j g_j(x)$ to a function $f(x)$:
  (i) Specify $n$ basis functions: $g(x, j) \equiv g_j(x)$ $(j = 1, \ldots, n)$
    Specify $n$ collocation points: $x = x_i$ $(i = 1, \ldots, n)$
    Specify the function $f(x)$, interval $[a, b]$, and spacing $h$ of comparison points.
  (ii) Solve $\mathbf{Ac} = \mathbf{f}$ for parameters $\mathbf{c} = (c_1, \ldots, c_n)^T$,
    where $\mathbf{A}, \mathbf{c}, \mathbf{f}$ are defined by (7.18), by Gauss elimination.

  (iii) Compare $f(x)$ and the approximation $\sum_{j=1}^{n} c_j g_j(x)$ at points $x$ spaced $h$ apart in $[a, b]$.

# Program 7.1 COLLOC: Generalised approximation by collocation

```
LIST
COLLOC

10    REM- COLLOC: FINDS GENERALISED APPROXN F*(X)=C(1)G(X,1)+...
20    REM- +C(N)G(X,N) BY COLLOCATION TO F(X) AT X(1),...,X(N).USES
30    REM- MATRIX ROUTINES TO SOLVE LINEAR ALGEBRAIC SYSTEM.
40    REM- COMPARES F* WITH F AT M EQUALLY SPACED POINTS.
50    DIM C(10,1),X(10),F(10,1),A(10,10)
60    PRINT "NUMBER OF PARAMETERS";
70    INPUT N
80    REM- REDIMENSIONS A,F,C APPROPRIATELY
90    MAT A=ZER(N,N)
100   MAT F=ZER(N,1)
110   MAT C=ZER(N,1)
120   PRINT "COLLOCATION POINTS X(1),...,X(N)"
130   FOR I=1 TO N
140   INPUT X(I)
150   NEXT I
160   REM- SETS UP AND SOLVES COLLOCATION EQNS
170   REM- F(X),G(X,J) ARE DEFINED BY FUNCTIONS FNF(X),FNG(X,J)
180   FOR I=1 TO N
190   F(I,1)=FNF(X(I))
200   FOR J=1 TO N
210   A(I,J)=FNG(X(I),J)
220   NEXT J
230   NEXT I
240   MAT A=INV(A)
250   MAT C=A*F
260   PRINT "PARAMETERS C(1),...,C(N)"
270   FOR I=1 TO N
280   PRINT C(I,1)
290   NEXT I
300   PRINT
310   REM- COMPARES F(X) AND F*(X) AT M PTS IN [A1,B1]
320   PRINT "NO OF COMPARISON PTS (>1)";
330   INPUT M
340   PRINT "LIMITS OF RANGE OF COMPARISON A1,B1 :";
350   INPUT A1,B1
360   H=(B1-A1)/(M-1)
370   PRINT "X VALUE:","FUNCTION:","APPROXN:","ERROR:"
380   FOR I=1 TO M
390   X1=A1+(I-1)*H
400   D=FNF(X1)
410   D1=0
420   FOR J=1 TO N
430   D1=D1+C(J,1)*FNG(X1,J)
440   NEXT J
450   E=D-D1
460   PRINT X1,D,D1,E
470   NEXT I
480   DEF FNF(X)=EXP(X)
490   DEF FNG(X,J)=X^(J-1)
500   END

Ready
```

## Sample run 1

```
RUN
COLLOC

NUMBER OF PARAMETERS? 4
COLLOCATION POINTS X(1),...,X(N)
? .125
? .375
? .625
? .875
PARAMETERS C(1),...,C(N)
 .998393
 1.02125
 .419741
 .276943
```

```
NO OF COMPARISON PTS (>1)? 21
LIMITS OF RANGE OF COMPARISON A1,B1 :? 0,1
X VALUE:      FUNCTION:      APPROXN:       ERROR:
  0            :             .998393         .160718E-02
  .05          1.05127       1.05054         .731707E-03
  .1           1.10517       1.10499         .178695E-03
  .15          1.16183       1.16196        -.12517E-03
  .2           1.2214        1.22165        -.245571E-03
  .25          1.28403       1.28427        -.241399E-03
  .3           1.34986       1.35002        -.163794E-03
  .35          1.41907       1.41912        -.555515E-04
  .4           1.49182       1.49178         .48399E-04
  .45          1.56831       1.56819         .12207E-03
  .5           1.64872       1.64857         .149488E-03
  .55          1.73325       1.73313         .123739E-03
  .6           1.82212       1.82207         .486374E-04
  .65          1.91554       1.9156         -.616312E-04
  .7           2.01375       2.01393        -.180721E-03
  .75          2.117         2.11727        -.271082E-03
  .8           2.22554       2.22582        -.282288E-03
  .85          2.33965       2.3398         -.150442E-03
  .9           2.4596        2.4594          .201941E-03
  .95          2.58571       2.58484         .867605E-03
  1            2.71828       2.71633         .195336E-02
Ready
```

## Sample run 2

```
RUN
COLLOC

NUMBER OF PARAMETERS? 4
COLLOCATION POINTS X(1),....,X(N)
? .03806
? .30866
? .69134
? .96194
PARAMETERS C(1),...,C(N)
 .999509
1.01563
 .424303
 .278241


NO OF COMPARISON PTS (>1)? 21
LIMITS OF RANGE OF COMPARISON A1,B1 :? 0,1
X VALUE:      FUNCTION:      APPROXN:       ERROR:
  0            1             .999509         .491261E-03
  .05          1.05127       1.05139        -.114918E-03
  .1           1.10517       1.10559        -.422359E-03
  .15          1.16183       1.16234        -.505447E-03
  .2           1.2214        1.22183        -.430584E-03
  .25          1.28403       1.28428        -.258207E-03
  .3           1.34986       1.3499         -.399351E-04
  .35          1.41907       1.41889         .180125E-03
  .4           1.49182       1.49146         .366569E-03
  .45          1.56831       1.56782         .492096E-03
  .5           1.64872       1.64818         .53978E-03
  .55          1.73325       1.73275         .501633E-03
  .6           1.82212       1.82174         .380635E-03
  .65          1.91554       1.91535         .19025BE-03
  .7           2.01375       2.0138         -.448227E-04
  .75          2.117         2.11729        -.287294E-03
  .8           2.22554       2.22603        -.488043E-03
  .85          2.33965       2.34023        -.584364E-03
  .9           2.4596        2.4601         -.499487E-03
  .95          2.58571       2.58585        -.141859E-03
  1            2.71828       2.71769         .595093E-03
Ready
```

*Sample run 3*

```
RUN COLLOC
Ready

480    DEF FNF(X)=X*(1-X*X)
490    DEF FNG(X,J)=SIN(J*PI*X)
RUN
COLLOC

NUMBER OF PARAMETERS? 3
COLLOCATION POINTS X(1),....,X(N)
? .25
? .5
? .75
PARAMETERS C(1),....,C(N)
 .386374
-.046875
 .113738E-01

NO OF COMPARISON PTS (>1)? 21
LIMITS OF RANGE OF COMPARISON A1,B1 :? 0,1
X VALUE:       FUNCTION:      APPROXN:       ERROR:
 0              0              0              0
 .05            .049875        .511206E-01   -.124561E-02
 .1             .099           .101045       -.204524E-02
 .15            .146625        .148721       -.209615E-02
 .2             .192           .193341       -.134116E-02
 .25            .234375        .234375       -.149012E-07
 .3             .273           .271517        .148311E-02
 .35            .307125        .30456         .256538E-02
 .4             .336           .333226        .277448E-02
 .45            .358875        .356998        .187743E-02
 .5             .375           .375           .298023E-07
 .55            .383625        .385968       -.234288E-02
 .6             .384           .38833        -.433034E-02
 .65            .375375        .380405       -.502992E-02
 .7             .357           .360678       -.367838E-02
 .75            .328125        .328125        .596046E-07
 .8             .288           .282503        .549731E-02
 .85            .235875        .224566        .113086E-01
 .9             .171           .15615         .01485
 .95            .092625        .800909E-01    .125341E-01
 1              0             -.42245E-07     .42245E-07
Ready
```

*Program notes*

(1) On the VAX system used, dynamic dimensioning is not available and so the correct dimensions for matrices **A, f, c** are defined implicitly by setting them to zero matrices of appropriate dimensions.

If dynamic dimensioning is available, then the following modifications may be made:

Delete 50, 80–110
90 DIM C(N, 1), X(N), F(N, 1), A(N, N).

(2) If the BASIC matrix inversion routines are unavailable, then the appropriate part of one of the programs of Chapter 5 (e.g. Program 5.3, lines 120–560, omitting REM statements and lines 450–460), might be used, instead of instructions 240–250, to solve **Ac** = **f**. Note, however, that the additional statement

DIM A1(10, 10), B1(10), R(10)

would be needed, and the variables X(I) and B(I) in Program 5.3

would have to be replaced by C(I, 1) and F(I, 1) throughout (where I is any index). All instructions would also have to be renumbered sequentially.

(3) The functions $f(x) = e^x$ and $g(x, j) = x^{j-1}$ are at present specified in instructions 480, 490. New instructions 480 or 490 must be input if a new $f(x)$ or $g(x, j)$ is required, as in Sample run 3.

(4) In Sample runs 1 and 2, the function $f(x) = e^x$ is fitted by a cubic on [0, 1] at equally spaced points and Chebyshev zeros, respectively. Note that the approximation is more accurate when the Chebyshev zeros are used. Moreover in this case the error oscillates with nearly equal extrema ($\cong$ .00049, .00051, .00054, .00058, .00060), and so the approximation is very close to *best*.

(5) In Sample run 3, the function $x(1 - x^2)$ is collocated on [0, 1] by the sum $c_1 \sin \pi x + c_2 \sin 2\pi x + c_3 \sin 3\pi x$. (The symbol PI in the new instruction 490 represents $\pi$ in VAX BASIC.)

It is necessary to point out, when offering the general algorithm 7.1, that there are two important choices left to the reader: the choice of basis functions, and the choice of collocation points. Even after deciding on, say, polynomials, the choice $\{g_j(x)\}$ is not unique. More importantly, different choices of collocation points lead to different approximations, and it is important to make a choice which is backed by theory. We must point out that, as collocation points for polynomial approximation, *a set of equally spaced points is not generally to be recommended* (except possibly for small values of $n$); indeed $f^*(x)$ can *diverge* from $f(x)$ for large $n$ (see Problem 6). In contrast Chebyshev polynomial zeros, as used in Sample run 2, are always a good choice of collocation points for a continuous function $f(x)$.

These points are amplified in the following subsections, and indeed a very efficient and reliable algorithm is given for polynomial collocation at Chebyshev polynomial zeros.

### 7.4.1  *Choice of basis functions*

Once a form of approximation $f^*(x)$ has been decided upon, there is still a variety of possible choices of basis functions. For example, although a polynomial of degree $n - 1$ can obviously be expressed in the form

$$f^*(x) = c_1 + c_2 x + \ldots + c_i x^{i-1} + \ldots + c_n x^{n-1} \tag{7.20}$$

it can also be expressed in terms of Chebyshev polynomials in the form

$$f^*(x) = c_1 T_0(x) + c_2 T_1(x) + \ldots + c_i T_{i-1}(x) + \ldots + c_n T_{n-1}(x) \tag{7.21}$$

where the parameters $c_1, \ldots, c_n$ take appropriate values (but not the same values in (7.21) as in (7.20)). For example, the quadratic $(1 + x)^2$ can be written in either of the forms

$$1 + 2x + x^2 \quad \text{or} \quad 1.5T_0(x) + 2T_1(x) + 0.5T_2(x)$$

The choice of basis functions has a considerable effect on the computation. Indeed the linear system (7.17) is rather ill-conditioned if the basis of $x$ powers (7.20) is used, but very well-conditioned if the basis of Chebyshev polynomials (7.21) is used.

### 7.4.2 Efficient polynomial collocation, Chebyshev zeros

For approximation by a polynomial of degree $n - 1$, we have already recommended the use of the Chebyshev polynomial zeros (7.10) appropriate to the range $[a, b]$ as collocation points. Indeed judging by Sample run 2 of Program 7.1, it would appear that collocation at these points leads to 'near-best' approximations. Why is this?

In the case of the 'model' function $f(x) \equiv x^n$ on $[-1, 1]$, the best (Chebyshev norm) approximation $f^*(x)$ of the polynomial form (7.20) (or equivalently (7.21)) is given by

$$f(x) - f^*(x) = 2^{1-n}T_n(x)$$

Thus $f(x) = f^*(x)$ at the $n$ zeros of $T_n(x)$, and so the polynomial $f^*(x)$ which collocates $f(x)$ at the Chebyshev zeros is the best approximation. By deeper arguments (see Reference 4) it can be proved that, for *any* continuous $f(x)$, the polynomial collocating at Chebyshev zeros is close to a best approximation.

Adopting the form (7.21), namely

$$f^*(x) = \sum_{j=1}^{n} c_j T_{j-1}(z), \quad \text{where } z = \frac{2}{b-a}\left\{x - \frac{a+b}{2}\right\} \quad (7.22)$$

the polynomial collocating at Chebyshev zeros appropriate to the range $[a, b]$ is determined explicitly by the following simple and efficient algorithm.

### Algorithm 7.2 Polynomial approximation by Chebyshev collocation

(i) Define $f(x)$ and $[a, b]$.
(ii) Calculate $x_i (i = 1, \ldots, n)$ from (7.10)
    and let $z_i$ be the values of $z$ corresponding to $x = x_i$ (from (7.22)).
(iii) For $j = 1, \ldots, n$:

define $c_j = \frac{w}{n} \sum_{i=1}^{n} f(x_i)T_{j-1}(z_i)$ \hfill (7.23) where $w = 1$ for $j = 1$ and $w = 2$ for $j > 1$.

(iv) Hence define $f^*(x)$ as the polynomial (7.22).

(v) Compare $f(x)$ and $f^*(x)$ at sample points $x$ spaced by $h$ in $[a, b]$.

This algorithm, and specifically formula (7.23), is based on the observation that the Chebyshev polynomials satisfy a 'discrete orthogonality' property (Reference 4). For $a = -1$, $b = 1$ (when $x_i = z_i$) this takes the form

$$\frac{2}{n} \sum_{i=1}^{n} T_{j-1}(x_i) T_{k-1}(x_i) = \begin{cases} 0 \text{ for } j \neq k \\ 2 \text{ for } j = k = 1 \\ 1 \text{ for } j = k > 1 \end{cases} \quad (7.24)$$

Now if we set $f(x) = f^*(x)$ at the points $x_i$, then

$$f(x_i) = \sum_{k=1}^{n} c_k T_{k-1}(x_i)$$

Hence $\quad \dfrac{2}{n} \sum_{i=1}^{n} f(x_i) T_{j-1}(x_i) = \sum_{k=1}^{n} c_k \left\{ \dfrac{2}{n} \sum_{i=1}^{n} T_{j-1}(x_i) T_{k-1}(x_i) \right\}$

$$= \begin{cases} 2c_j \text{ for } j = 1 \\ c_j \text{ for } j > 1 \end{cases} \text{ from } (7.24)$$

Thus (7.23) follows.

The following program implements this algorithm.

**Program 7.2** CHCOLL: Polynomial approximation by Chebyshev collocation

```
LIST
CHCOLL

10      REM- CHCOLL: FINDS APPROXN F*(X) AS SUM OF CJ*T(J-1)(Z) (J=1,...,N)
20      REM- ON RANGE [A,B] OF X, WHERE X=.5(A+B)+.5(B-A)Z, BY COLLOCATION
30      REM- TO GIVEN FUNCTION F(X) AT N ZEROS OF CHEBYSHEV POLY TN(Z).
40      DIM X(50),C(50)
50      PRINT "NO OF PARAMETERS";
60      INPUT N
70      PRINT "LIMITS OF RANGE OF APPROXN : A,B"
80      INPUT A,B
90      REM- CALCULATE CHEBYSHEV ZEOS
100     FOR I=1 TO N
110     X(I)=.5*((A+B)+(B-A)*COS((I-.5)*PI/N))
120     NEXT I
130     REM- CALCULATE APPROXIMATION COEFFTS CJ
140     PRINT "J","C(J)"
150     FOR J=1 TO N
160     D=0
170     FOR I=1 TO N
180     D=D+FNF(X(I))*COS((J-1)*(I-.5)*PI/N)
190     NEXT I
200     IF J>1 THEN 230
210     C(1)=D/N
220     GO TO 240
230     C(J)=2*D/N
240     PRINT J,C(J)
250     NEXT J
260     PRINT "NO OF COMPARISON PTS (>1)";
```

```
270     INPUT M
280     H1=(B-A)/(M-1)
290     H=2/(M-1)
300     REM- COMPARES F(X) AND F*(X) AT M PTS IN [A,B]
310     PRINT "X VALUE:","FUNCTION:","APPROXN:","ERROR:"
320     FOR I=1 TO M
330     X1=A+(I-1)*H1
340     Z=-1+(I-1)*H
350     D=FNF(X1)
360     D1=0
370     REM- CALCULATES V=ARCCOS(Z) IN RANGE [0,PI]
380     IF Z<>0 THEN 410
390     V=PI/2
400     GO TO 450
410     V=ATN(SQR(1-Z*Z)/Z)
420     IF Z>0 THEN 450
430     V=PI+V
440     REM- CALCULATES D1=F*(X)=SUM OF CJ*T(J-1)(Z)
450     FOR J=1 TO N
460     D1=D1+C(J)*COS((J-1)*V)
470     NEXT J
480     E=D-D1
490     PRINT X1,D,D1,E
500     NEXT I
510     REM- DEFINES F(X) AS FNF(X)
520     DEF FNF(X)=EXP(X)
530     END
```

## Sample run

```
RUN
CHCOLL

NO OF PARAMETERS? 4
LIMITS OF RANGE OF APPROXN : A,B
? 0,1
J               C(J)
 1              1.75339
 2               .850392
 3               .105208
 4               .86948E-02
NO OF COMPARISON PTS (>1)? 21
X VALUE:    FUNCTION:       APPROXN:        ERROR:
 0          1               .999509          .490785E-03
 .05        1.05127         1.05139         -.115156E-03
 .1         1.10517         1.10559         -.422239E-03
 .15        1.16183         1.16234         -.504851E-03
 .2         1.2214          1.22183         -.430107E-03
 .25        1.28403         1.28428         -.257373E-03
 .3         1.34986         1.3499          -.389814E-04
 .35        1.41907         1.41889          .181437E-03
 .4         1.49182         1.49146          .367999E-03
 .45        1.56831         1.56782          .493646E-03
 .5         1.64872         1.64818          .541329E-03
 .55        1.73325         1.73275          .503421E-03
 .6         1.82212         1.82174          .382662E-03
 .65        1.91554         1.91535          .192404E-03
 .7         2.01375         2.0138          -.424385E-04
 .75        2.117           2.11728         -.284433E-03
 .8         2.22554         2.22603         -.485182E-03
 .85        2.33965         2.34023         -.581264E-03
 .9         2.4596          2.4601          -.495672E-03
 .95        2.58571         2.58585         -.137329E-03
 1          2.71828         2.71768          .599623E-03
Ready
```

## Program notes

(1) $T_{j-1}(z)$ is calculated as $\cos(j-1)v$, where $v = \cos^{-1} z$, in instructions 380–430, 460. Note that, since the function $\cos^{-1}$ is not available in BASIC, we have used the function $\tan^{-1}\{\sqrt{(1-z^2)}/z\}$ adjusted

to give a value in $[0, \pi]$. (PI represents $\pi$ throughout the program.)
(2) The calculation of $T_{j-1}(z)$ via $\cos(j-1)v$ is simple but inefficient.
An efficient algorithm for calculating $\Sigma c_j T_{j-1}(z)$, based on a recurrence, is given by Cheney (Reference 4).
(3) Note that virtually identical results are obtained (as they should be) in this Sample run to those in run 2 of Program 7.1.

## 7.5  Data fitting by least squares

So far we have mainly been concerned with approximation throughout an interval $[a, b]$, but now we turn our attention to the approximation of a discrete set of data subject to experimental error. There is then really only one technique which is versatile and elementary, and that is the method of least squares which provides an 'unbiased assessment' of an appropriate fit.

In data fitting it is not uncommon to use a very simple approximating curve, such as a straight line or parabola, and indeed there are a number of situations in which this is reasonable. Let us therefore start with a straight line, for which explicit results may be calculated easily.

### 7.5.1  Straight line fit

Assume that the independent variable $x$ is exact, while the dependent variable $y = f(x)$ has random errors occurring in a normal distribution. Suppose that there are $m$ data points $(x_i, y_i)$ $(i = 1, \ldots, m)$, and that values of $c_1$ and $c_2$ are required to define a straight line approximation

$$y \simeq f^*(x) = c_1 + c_2 x \qquad (7.25)$$

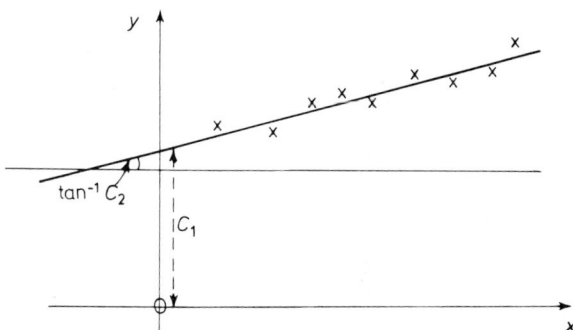

*Figure* 7.5. Least squares straight line fit.

where $c_1$ is the intercept with the $y$ axis, and $c_2$ is the gradient. (See Figure 7.5.) Adopting the *least squares norm* (7.2), namely

$$s_1 = \|f - f^*\| = \left[\frac{1}{m} \sum_{i=1}^{m} \{f(x_i) - f^*(x_i)\}^2\right]^{1/2}$$

$$= \left[\frac{1}{m} \sum (y_i - c_1 - c_2 x_i)^2\right]^{1/2} \qquad (7.26)$$

we seek the best approximation of form (7.25) which minimises $s_1$.

Clearly the minimum of $s_1$ coincides with the minimum of $ms_1^2$, that is of

$$E = \sum_{i=1}^{m} (y_i - c_1 - c_2 x_i)^2$$

This sum has an extreme value when

$$\frac{\partial E}{\partial c_1} = 0 \quad \text{and} \quad \frac{\partial E}{\partial c_2} = 0$$

Now

$$\frac{\partial E}{\partial c_1} = -2 \sum (y_i - c_1 - c_2 x_i) = 0$$

and so

$$c_1(m) + c_2 \sum x_i = \sum y_i \qquad (7.27)$$

where the sums are over $i = 1, \ldots, m$.

Also

$$\frac{\partial E}{\partial c_2} = -2 \sum x_i(y_i - c_1 - c_2 x_i) = 0$$

and so

$$c_1 \sum x_i + c_2 \sum x_i^2 = \sum x_i y_i \qquad (7.28)$$

Solving (7.27) and (7.28), $c_1$ and $c_2$ are determined uniquely as

$$c_2 = \frac{m \sum x_i y_i - \sum x_i \sum y_i}{m \sum x_i^2 - (\sum x_i)^2}, \quad c_1 = \frac{\sum y_i - c_2 \sum x_i}{m} \qquad (7.29)$$

Note that the line of best fit $y = c_1 + c_2 x$ always passes through the centre of gravity of the data points, since

$$\bar{x} = \sum x_i / m, \quad \bar{y} = \sum y_i / m \quad \text{and} \quad c_1 = \bar{y} - c_2 \bar{x}$$

The root mean square error (7.26) is an approximation to the 'standard deviation' of the fit defined by

$$\sigma = \left[\frac{1}{m} \sum (y_i^{\mathrm{T}} - c_1 - c_2 x_i)^2\right]^{1/2} \qquad (7.30)$$

where $y_i^{\mathrm{T}}$ is the true value at $x_i$ of the function underlying the experimental data. However, statistical theory shows that a more realistic estimate of (7.30) (in which $y_i^T$ is unknown) is given by the *estimated standard deviation*

$$s_2 = \left[ \frac{1}{m-2} \sum (y_i - c_1 - c_2 x_i)^2 \right]^{1/2} \tag{7.31}$$

The denominator $m - 2$ in the square root, which replaces the denominator $m$ in the r.m.s error (7.26), measures the number of 'degrees of freedom' in the data (i.e. $m$ data less 2 parameters).

The following algorithm and program implement the complete method.

*Algorithm 7.3 Least squares straight line fit*

(i) Specify data $(x_i, y_i)$ $(i = 1, \ldots, m)$.
(ii) Calculate the centre of gravity $(\bar{x}, \bar{y})$.
(iii) Calculate $c_1, c_2$ (to define the straight line) from (7.29).
(iv) Calculate the r.m.s. error $s_1$ from (7.26) and the estimated standard deviation $s_2$ from (7.31).

**Program 7.3** LINLSQ: Least squares straight line fit

```
LIST
LINLSQ

10      REM- LEAST SQUARES FIT OF ST LINE Y=C1+C2*X TO DATA X,Y
20      DIM X(50),Y(50)
30      PRINT "NO OF DATA PTS";
40      INPUT M
50      REM- D,E,F,G BECOME SUMS OF X,Y,X*Y,X*X
60      D=0
70      E=0
80      F=0
90      G=0
100     PRINT "DATA X,Y:"
110     FOR I=1 TO M
120     INPUT X(I),Y(I)
130     D=D+X(I)
140     E=E+Y(I)
150     F=F+X(I)*Y(I)
160     G=G+X(I)*X(I)
170     NEXT I
180     XO=D/M
190     YO=E/M
200     PRINT "X,Y COORDS OF CENTRE OF GRAVITY:";XO;",";YO
210     C2=(M*F-D*E)/(M*G-D*D)
220     C1=YO-C2*XO
230     PRINT "COEFFICIENTS C1:";C1;",C2:";C2
240     S=0
250     PRINT "X :","GIVEN Y :","FITTED Y :","ERROR :"
260     FOR I=1 TO M
270     Y1=C1+C2*X(I)
280     E=Y(I)-Y1
290     PRINT X(I),Y(I),Y1,E
300     S=S+E*E
310     NEXT I
320     S1=SQR(S/M)
330     S2=SQR(S/(M-2))
340     PRINT "R.M.S. ERROR:",S1
350     PRINT "ESTIMATED STANDARD DEVIATION:";S2
360     END

Ready
```

## Sample runs

```
RUN
LINLSQ

NO OF DATA PTS? 5
DATA X,Y:
? 0,0
? 1,1.1
? 2,1.1
? 3,3.4
? 4,3.9
X,Y COORDS OF CENTRE OF GRAVITY: 2 , 1.9
COEFFICIENTS C1:-.12 ,C2: 1.01
X :          GIVEN Y :      FITTED Y :      ERROR :
  0             0              -.12           .12
  1            1.1             .89            .21
  2            1.1             1.9           -.8
  3            3.4             2.91           .49
  4            3.9             3.92          -.200005E-01
R.M.S. ERROR:   .433359
ESTIMATED STANDARD DEVIATION: .559464
Ready
```

```
RUN
LINLSQ

NO OF DATA PTS? 5
DATA X,Y:
? 0,0
? 1,1.1
? 2,2.1
? 3,3.4
? 4,3.9
X,Y COORDS OF CENTRE OF GRAVITY: 2 , 2.1
COEFFICIENTS C1: .799999E-01 ,C2: 1.01
X :          GIVEN Y :      FITTED Y :      ERROR :
  0             0             .799999E-01   -.799999E-01
  1            1.1            1.09           .999999E-02
  2            2.1            2.1           -.238419E-06
  3            3.4            3.11           .29
  4            3.9            4.12          -.22
R.M.S. ERROR:   .166733
ESTIMATED STANDARD DEVIATION: .215252
Ready
```

## Program notes

(1) In Sample run 1, the third data point is apparently wild, and in consequence a much larger r.m.s. error is obtained than in Sample run 2. Indeed the mean error given by (7.3), would probably be a better norm to choose for fitting the first data set (but that would require another algorithm).

Before considering a more sophisticated fit it is useful to point out that, even when data are not well fitted by a straight line, it is often possible to *rectify* (or transform) the data into a form which does approximate a straight line, by using logarithms, reciprocals, etc. For example, the law

$$y = Ax^B$$

is equivalent to

$$\log y = \log A + B(\log x)$$

and so a straight line may be fitted (where logs may be taken to either base e or base 10) to the data of $\log y$ against $\log x$. However, it is important to note that, after obtaining acceptable errors in fitting $\log y$, we might subsequently produce large errors in $y = \exp(\log y)$ when we transform back to the original variables $x$, $y$!

### 7.5.2 Generalised fit

The method used above to determine a straight line fit can easily be extended to determine a generalised linear approximation (as in Algorithm 7.1) of the form

$$f^*(x) = c_1 g_1(x) + c_2 g_2(x) + \ldots + c_n g_n(x) = \sum_{j=1}^{n} c_j g_j(x) \quad (7.32)$$

where any number $n$ of basis functions may be used and any choice of $g_1(x), \ldots, g_n(x)$ may be made (provided that $g_1, \ldots, g_n$ are linearly independent).

The case of a polynomial fit of degree $n - 1$ corresponds to the obvious choice

$$g_j(x) = x^{j-1} \quad (7.33)$$

(although this is not the only possible choice of basis functions in this case) and a straight line fit is obtained from (7.33) for $n = 2$.

For a set of $m$ data points $(x_i, y_i)$ $(i = 1, \ldots, m)$, where $m \geqslant n$, the least squares norm (r.m.s error) for (7.32) is

$$s_1 = \| f - f^* \| = \left[ \frac{1}{m} \sum_{i=1}^{m} \left\{ y_i - \sum_{j=1}^{n} c_j g_j(x_i) \right\}^2 \right]^{1/2} \quad (7.34)$$

and the best approximation is obtained by minimising with respect to each $c_k$ $(k = 1, \ldots, n)$ the quantity

$$E = m s_1^2 = \sum_{i=1}^{m} \left\{ y_i - \sum_{j=1}^{n} c_j g_j(x_i) \right\}^2$$

The minimum is given by the set of equations

$$\frac{\partial E}{\partial c_k} = \sum_{i=1}^{m} 2 \left\{ y_i - \sum_{j=1}^{n} c_j g_j(x_i) \right\} \{ -g_k(x_i) \} = 0 \quad (k = 1, \ldots, n)$$

Thus $\displaystyle \sum_{j=1}^{n} c_j \left\{ \sum_{i=1}^{m} g_j(x_i) g_k(x_i) \right\} = \sum_{i=1}^{m} y_i g_k(x_i) \quad (k = 1, \ldots, n) \quad (7.35)$

which is a system of $n$ linear equations for $c_1, \ldots, c_n$. We may use any of the algorithms of Chapter 5 to solve (7.35) and hence determine the fit $f^*(x)$. Indeed the matrix on the left hand is symmetric and positive definite, and so Program 5.5 is suitable.

Note that (7.35) may be written in matrix form as

$$\mathbf{Ac} = \mathbf{b} \tag{7.36}$$

where $\mathbf{A}$ has elements $A_{kj}$ (in row $k$, column $j$), and $\mathbf{c}$ and $\mathbf{b}$ have elements $c_j$ and $b_k$ (in rows $j$ and $k$, respectively) for $j = 1, \ldots, n$ and $k = 1, \ldots, n$.

Now
$$A_{kj} = \sum_{i=1}^{m} g_j(x_i)g_k(x_i), \quad b_k = \sum_{i=1}^{m} y_i g_k(x_i)$$

and, since $A_{jk} = A_{kj}$, $\mathbf{A}$ is clearly symmetric.

Also
$$\mathbf{A} = \mathbf{B}^{\mathrm{T}}\mathbf{B} \quad \text{and} \quad \mathbf{b} = \mathbf{B}^{\mathrm{T}}\mathbf{y} \tag{7.37}$$

where $B_{ij} = g_j(x_i)$ and $y_i$ are data values. $(i = 1, \ldots, m; j = 1, \ldots, n)$. So the system (7.36) is effectively generated by multiplying through by $\mathbf{B}^{\mathrm{T}}$ in the linear system

$$\mathbf{Bc} = \mathbf{y} \tag{7.38}$$

The latter is exactly the set of equations for $\mathbf{c}$ which are obtained by equating $f^*(x_i)$ to $y_i$ at each of the $m$ data points. Clearly (7.38) cannot generally be solved (for $m > n$), since it is an *over-determined* set of $m$ equations in $n$ unknowns. Indeed, in minimising $\|f - f^*\|_2$ by solving (7.36), we are solving the over-determined system (7.38) approximately *in the least squares sense*. (Equations (7.36) are called the *normal equations* for (7.38), and the symmetric matrix $\mathbf{A} = \mathbf{B}^{\mathrm{T}}\mathbf{B}$ is called the *normal matrix* of the rectangular matrix $\mathbf{B}$.)

The estimated standard deviation for $f^*(x)$, based on $m - n$ degrees of freedom, is

$$s_2 = \left\{ \frac{1}{m-n} \sum_{i=1}^{m} \left[ y_i - \sum_{j=1}^{n} c_j g_j(x_i) \right]^2 \right\}^{1/2} \tag{7.39}$$

In the case $m = n$, (7.38) can be solved exactly, and indeed $f^*(x)$ coincides with the collocation polynomial at $x_1, \ldots, x_n$. The r.m.s. error is then zero, and $s_2$ is not defined.

The above method is now implemented in an algorithm, and, as in Section 7.4, a function $g(x, j)$ of 2 variables is used to define $g_j(x)$.

### Algorithm 7.4 Generalised least squares data fitting

(i) Specify the data $(x_i, y_i)$ $(i = 1, \ldots, m)$.
(ii) Specify $n$ basis functions $g(x, j) = g_j(x)$.

(iii)  Solve (7.36) for parameters $\mathbf{c} = (c_1, \ldots, c_n)^T$
where $\mathbf{A}$ and $\mathbf{b}$ are defined by (7.37).

(iv)  Calculate the r.m.s. error $s_1$ from (7.34) and estimated standard
deviation $s_2$ from (7.39).

## Program 7.4 LEASQU: Generalised least squares fit

```
LIST
LEASQU

10      REM- LEASQU: FITS C(1)G(X,1)+...+C(N)G(X,N) TO Y VALUES AT
20      REM- X(1),...,X(M) BY LEAST SQUARES, USING BASIC MATRIX ROUTINES
30      DIM X(50),Y(50,1),B(50,10),B1(10,50),A(10,10),C(10,1),U(10,1)
40      PRINT "NO OF DATA PTS";
50      INPUT M
60      PRINT "X,Y VALUES:"
70      REM- REDIMENSION Y
80      MAT Y=ZER(M,1)
90      FOR I=1 TO M
100       INPUT X(I),Y(I,1)
110     NEXT I
120     PRINT "NO OF APPROXN PARAMETERS";
130     INPUT N
140     REM- REDIMENSION B,B1,A,C,U
150     MAT B=ZER(M,N)
160     MAT B1=ZER(N,M)
170     MAT A=ZER(N,N)
180     MAT C=ZER(N,1)
190     MAT U=ZER(N,1)
200     REM- FORMS MATRIX B OF OVERDETERMINED EQNS BC=Y
210     REM- G(X,J) IS DEFINED BY FUNCTION FNG(X,J)
220     FOR I=1 TO M
230     FOR J=1 TO N
240     B(I,J)=FNG(X(I),J)
250     NEXT J
260     NEXT I
270     MAT B1=TRN(B)
280     REM- SOLVE NORMAL EQNS (B'B)C=B'Y=U
290     MAT A=B1*B
300     MAT U=B1*Y
310     MAT A=INV(A)
320     MAT C=A*U
330     PRINT "PARAMETERS C(1),...,C(N)"
340     FOR I=1 TO N
350     PRINT C(I,1)
360     NEXT I
370     PRINT
380     REM- COMPARE APPROXN C(1)G(X,1)+... WITH Y AT DATA PTS
390     PRINT "X :","GIVEN Y :","FITTED Y :","ERROR :"
400     R=0
410     FOR I=1 TO M
420     D=Y(I,1)
430     D1=0
440     FOR J=1 TO N
450     D1=D1+C(J,1)*B(I,J)
460     NEXT J
470     E=D-D1
480     R=E*E+R
490     PRINT X(I),D,D1,E
500     NEXT I
510     S1=SQR(R/M)
520     PRINT "R.M.S. ERROR :";S1
530     IF M=N THEN 560
540     S2=SQR(R/(M-N))
550     PRINT "ESTIMATED STANDARD DEVIATION :";S2
560     DEF FNG(X,J)=X^(J-1)
570     END

Ready
```

## Sample run 1

```
RUN
LEASQU

NO OF DATA PTS? 9
X,Y VALUES:
? 0,0
? 1,.5
? 3,1
? 4,.866O
? 6,0
? 7,-.5
? 9,-1
? 10,-.866O
? 12,0
NO OF APPROXN PARAMETERS? 4
PARAMETERS C(1),...,C(N)
-.636034E-01
 .897451
-.220991
 .012234

X :            GIVEN Y :      FITTED Y :      ERROR :
 0              0             -.636034E-01     .636034E-01
 1              .5             .625091        -.125091
 3              1              .970147         .298535E-01
 4              .866           .773316         .926836E-01
 6              0              .796032E-02    -.796032E-02
 7             -.5            -.413757        -.862427E-01
 9             -1             -.968255        -.031745
 10            -.866          -.954225         .882246E-01
 12             0              .234127E-01    -.234127E-01
R.M.S. ERROR : .715038E-01
ESTIMATED STANDARD DEVIATION : .959325E-01
Ready
```

## Sample run 2

```
OLD LEASQU
Ready



560    DEF FNG(X,J)=SIN(J*PI*X)


RUN
LEASQU

NO OF DATA PTS? 6
X,Y VALUES:
? 0,0
? .2,.192
? .4,.336
? .6,.384
? .8,.288
? 1,0
NO OF APPROXN PARAMETERS? 3
PARAMETERS C(1),...,C(N)
 .386759
-.478061E-01
 .133207E-01

X :            GIVEN Y :      FITTED Y :      ERROR :
 0              0              0               0
 .2             .192           .194534        -.253376E-02
 .4             .336           .3319           .409967E-02
 .6             .384           .3881          -.40997E-02
 .8             .288           .285466         .25337E-02
 1              0             -.424879E-07     .424879E-07
R.M.S. ERROR : .278252E-02
ESTIMATED STANDARD DEVIATION : .393507E-02
Ready
```

*Program notes*

(1) Dynamic dimensioning has been achieved implicitly, but it may be done explicitly if the facility is available (see Note 1 on Program 7.1).

(2) The BASIC matrix inversion routines, used here to solve (7.36) for **c** may be replaced by the appropriate part of Program 5.5 with the variables X(I) and B(I) replaced by C(I, 1) and U(I, 1).

(3) The function $g(x, j) = x^{j-1}$ is at present specified by the defined function FNG, which may be redefined if required (as in Sample run 2).

(4) In Sample run 1, a cubic polynomial is fitted to 9 exact data points in $[0, 12]$ of the function $\sin(\pi x/6)$, and in Sample run 2 exact data of the function $x(1 - x^2)$ are fitted in $[0, 1]$ by $c_1 \sin \pi x + c_2 \sin 2\pi x + c_3 \sin 3\pi x$. Note the similarities between the results of Sample run 2 and those of Sample run 3 of Program 7.1. (PI represents $\pi$.)

The reader is not advised to use the above algorithm for large values of $n$. In fact the system (7.36) is potentially even more ill-conditioned than the generalised collocation equations (7.16). For while the conditioning of the set of equations (7.38) is comparable with that of a set of collocation equations, the situation is worsened on multiplying by $\mathbf{B}^T$. Indeed in the simple case $m = n$ in which collocation coincides with least squares fitting, twice as many figures are lost in solving (7.36) as in solving (7.38).

However, for polynomial least squares fitting, it is possible to give an algorithm which is not only well-conditioned but also very much more efficient than Algorithm 7.4. Of course it is more complicated!

### 7.5.3 *An orthogonal polynomial algorithm*

Algorithm 7.4 would be greatly simplified if we could choose a set of basis functions $\{g_j(x)\}$ with the property that

$$\sum_{i=1}^{m} g_j(x_i)g_k(x_i) = A_{kj} = 0 \quad \text{for } j \neq k \tag{7.40}$$

and such basis functions are termed *orthogonal* with respect to the points $\{x_i\}$. The matrix **A** with entries $A_{kj}$ then becomes a diagonal matrix and equations (7.36) immediately give an explicit formula for $c_j$:

$$c_j = b_j/A_{jj} = \sum_{i=1}^{m} y_i g_j(x_i) / \sum_{i=1}^{m} [g_j(x_i)]^2 \quad (j = 1. \ldots, n) \tag{7.41}$$

In the case of polynomial approximation, we can in fact always satisfy (7.40) while choosing $\phi_j(x)$ to be a polynomial of degree $j - 1$. Efficient algorithms, both for determining $\phi_j(x)$ and for computing $f^*(x)$ in the form (7.32), are described by Cheney (Reference 4, Chap. 4: Theorem 2 and Corollary) and a practical account is given by Forsythe (Reference 5). Although the full details are beyond the scope of this book, we give a simple (but not most efficient) procedure, called the *Gram–Schmidt orthogonalisation procedure*, for defining $g_j(x)$.

Assume that $g_1(x), \ldots, g_n(x)$ may be defined successively as

$$g_1(x) = 1, \quad g_j(x) = x^{j-1} - \sum_{r=1}^{j-1} \alpha_r^{(j)} g_r(x) \quad (j = 2, \ldots, n) \quad (7.42)$$

where $\alpha_1^{(j)}, \ldots, \alpha_{j-1}^{(j)}$ $(j = 2, \ldots, n)$ are sets of constants.

Then it is obvious (by induction) that $g_j(x)$ is a polynomial of degree $j - 1$. Multiplying (7.42) by $g_k(x)$, setting $x = x_i$, and summing over $i = 1, \ldots, n$:

$$\sum_{i=1}^{m} g_k(x_i) g_j(x_i) = \sum_{i=1}^{m} g_k(x_i)(x_i)^{j-1} - \sum_{r=1}^{j-1} \alpha_r^{(j)} \sum_{i=1}^{m} g_k(x_i) g_r(x_i)$$

For any $k \neq j$ it follows by (7.40) that

$$0 = \sum_{i=1}^{m} g_k(x_i)(x_i)^{j-1} - \alpha_k^{(j)} \sum_{i=1}^{m} \{g_k(x_i)\}^2$$

and so

$$\alpha_k^{(j)} = \sum_{i=1}^{m} g_k(x_i)(x_i)^{j-1} \bigg/ \sum_{i=1}^{m} \{g_k(x_i)\}^2 \quad (7.43)$$

Thus the polynomials $\{g_j(x)\}$ are defined by (7.42), with $\alpha_k^{(j)}$ given by (7.43).

This means that the best least squares polynomial approximation may be calculated *explicitly* (i.e. without solving a set of equations) in the form (7.32) by using the formulae (7.42) and (7.43). We shall not present a BASIC program for an orthogonal polynomial algorithm, but will leave this as an exercise (Problem 12) for interested readers, who are referred to Reference 5.

### 7.6 Spline approximation to functions and data

A 'spline function' of degree $p - 1$, say, is a form of approximation, originally developed in engineering modelling, which has two important features. Firstly, the interval $[a, b]$ of interest (or an interval including all data points) is split up into a number of pieces and a polynomial approximation of degree $p - 1$ is formed on each piece. Such an approximation is in general called a *piecewise polynomial* approximation of degree $p - 1$. Secondly, this piecewise polynomial is chosen in such a way that, when all the pieces are joined together, the resulting function is continuous at all the joins (commonly called *knots*), and so also are its 1st, 2nd, ..., $(p - 2)$th derivatives.

Thus a cubic spline ($p = 4$) is a piecewise cubic polynomial which is continuous up to its 2nd derivative. Cubic splines are frequently adequate in practice, since the human eye is not usually able to 'spot the joins'. An example of a cubic spline on [0, 2], having one knot at $x = 1$, is the function

$$S(x) = \begin{cases} x^3 & \text{for } 0 \leqslant x < 1 \\ x^3 + (x-1)^3 & \text{for } 1 \leqslant x \leqslant 2 \end{cases}$$

The two pieces of this function are

$$S_0(x) = x^3 \quad \text{and} \quad S_1(x) = x^3 + (x-1)^3$$

Clearly the difference between these, namely

$$S_1(x) - S_0(x) = (x-1)^3$$

has the property that it vanishes at $x = 1$, as also do its 1st and 2nd derivatives. So $S_0(x)$ and $S_1(x)$ share the same value and the same first and second derivatives at $x = 1$ as required.

More generally, a cubic spline approximation having $n - 4$ knots at $x = z_1, \ldots, z_{n-4}$ (to give a total of $n$ parameters $c_1, \ldots, c_n$) has the form

$$f^*(x) \equiv S(x) = c_1 + c_2 x + c_3 x^2 + c_4 x^3 + c_5(x - z_1)_+^3 + \\ \ldots + c_n(x - z_{n-4})_+^3 \quad (7.44)$$

i.e.
$$f^*(x) \equiv S(x) = \sum_{j=1}^{4} c_j x^{j-1} + \sum_{k=1}^{n-4} c_{k+4}(x - z_k)_+^3 \quad (7.45)$$

where for any $a$,

$$(x - a)_+^3 = \begin{cases} (x-a)^3 & \text{for } x \geqslant a \\ 0 & \text{for } x \leqslant a \end{cases} \quad (7.46)$$

In other words the suffix $+$ means that $(x - z_k)^3$ is only included for points to the right of the knot $z_k$. The knots are assumed to occur in their natural order within the interval $[a, b]$ of approximation. Setting $z_0 = a$ and $z_{n-3} = b$, the spline (7.45) has pieces

$$S(x) = S_l(x), \text{ say, for } z_l \leqslant x \leqslant z_{l+1} \quad (l = 0, \ldots, n - 4) \quad (7.47)$$

where
$$S_0(x) = \sum_{j=1}^{4} c_j x^{j-1}$$

$$S_1(x) = \sum_{j=1}^{4} c_j x^{j-1} + c_5(x - z_1)^3 = S_0(x) + c_5(x - z_1)^3, \text{ etc.}$$

Generally
$$S_l(x) = \sum_{j=1}^{4} c_j x^{j-1} + \sum_{k=1}^{l} c_{k+4}(x - z_k)^3 \tag{7.48}$$

and
$$S_l(x) = S_{l-1}(x) + c_{l+4}(x - z_l)^3 \tag{7.49}$$

Thus (7.47), (7.48) give distinct definitions for $S(x)$ in the separate pieces $[z_l, z_{l+1}]$ of $[a, b]$ in terms of conventional cubic polynomials (i.e. without using the new notation (7.46)). However, (7.45) provides a more efficient 'total representation'.

Note from (7.49) that the difference $S_l(x) - S_{l-1}(x)$ is just $c_{l+4}(x - z_l)^3$. This cubic vanishes at $x = z_l$, as also do its first and second derivatives, and so $S_l(x)$ and $S_{l-1}(x)$ match as required at the knot $x = z_l$.

More generally a spline of degree $p - 1$ with knots at $z_1, \ldots, z_{n-p}$, would be defined by

$$f^*(x) = S(x) = \sum_{j=1}^{p} c_j x^{j-1} + \sum_{k=1}^{n-p} c_{k+p}(x - z_k)_+^{p-1} \tag{7.50}$$

or
$$S(x) = S_l(x) = \sum_{j=1}^{p} c_j x^{j-1} + \sum_{k=1}^{l} c_{k+p}(x - z_k)^{p-1} \tag{7.51}$$
$$\text{for } z_l \leqslant x \leqslant z_{l+1}$$

An approximation of form (7.50) may be obtained to a given function $f(x)$ or given data $\{y_i\}$ in many ways. We shall consider in particular the use of the generalised algorithms for collocation and least squares.

### 7.6.1 Spline collocation

Collocation by $f^*(x)$ in the form (7.45) can obviously be achieved with the generalised collocation Program 7.1 by choosing basis functions:

$$g_1(x), \ldots, g_n(x) = 1, x, x^2, x^3, (x - z_1)_+^3, \ldots, (x - z_{n-4})_+^3$$

i.e.    $g_j(x) = x^{j-1}$ $(j = 1, \ldots, 4)$; $g_j(x) = (x - z_{j-4})_+^3$
$$(j = 5, \ldots, n) \quad (7.52)$$

More generally for a spline of degree $p - 1$, given by (7.50), we would use

$$g_j(x) = x^{j-1} \ (j = 1, \ldots, p); \quad g_j(x) = (x - z_{j-p})_+^{p-1} \ (j = p+1, \ldots, n).$$

Note that this definition of $\{g_j(x)\}$ cannot be accomplished in one line of BASIC with $g(x, j) = g_j(x)$, and indeed it is clear that a BASIC function of $x$ and $j$ only cannot define $(x - z_{j-p})_+^{p-1}$.

However, suitable special modifications are simple to make, as the following Program and its Program Notes explain.

**Program 7.5** SPCOLL: Spline approximation by collocation

```
LIST
SPCOLL

10      REM- SPCOLL: SPECIALISATION OF "COLLOC".
20      REM- COLLOC: FINDS GENERALISED APPROXN F*(X)=C(1)G(X,1)+...
30      REM- +C(N)G(X,N) BY COLLOCATION TO F(X) AT X(1),...,X(N).USES
40      REM- MATRIX ROUTINES TO SOLVE LINEAR ALGEBRAIC SYSTEM.
50      REM- COMPARES F* WITH F AT M EQUALLY SPACED POINTS.
60      REM- F*(X)=SPLINE OF DEG (P-1) WITH KNOTS Z(1),...,Z(N-P).
70      DIM C(10,1),X(10),F(10,1),A(10,10)
80      DIM Z(10)
90      PRINT "NUMBER OF PARAMETERS";
100     INPUT N
110     REM- REDIMENSIONS A,F,C APPROPRIATELY
120     MAT A=ZER(N,N)
130     MAT F=ZER(N,1)
140     MAT C=ZER(N,1)
150     PRINT "DEGREE OF SPLINE PLUS 1";
160     INPUT P
170     PRINT N-P;"KNOTS OF SPLINE"
180     FOR I=1 TO N-P
190     INPUT Z(I)
200     NEXT I
210     PRINT "COLLOCATION POINTS X(1),...,X(N)"
220     FOR I=1 TO N
230     INPUT X(I)
240     NEXT I
250     REM- SETS UP AND SOLVES COLLOCATION EQNS
260     REM- F(X),G(X,J) ARE DEFINED BY FUNCTIONS FNF(X),FNG(X,J)
270     FOR I=1 TO N
280     F(I,1)=FNF(X(I))
290     FOR J=1 TO P
300     A(I,J)=FNG(X(I),J)
310     NEXT J
320     FOR J=P+1 TO N
330     IF Z(J-P) > X(I) THEN 350
340     A(I,J)=(X(I)-Z(J-P))^(P-1)
350     NEXT J
360     NEXT I
370     MAT A=INV(A)
380     MAT C=A*F
390     PRINT "PARAMETERS C(1),...,C(N)"
400     FOR I=1 TO N
410     PRINT C(I,1)
420     NEXT I
430     PRINT
440     REM- COMPARES F(X) AND F*(X) AT M PTS IN [A1,B1]
450     PRINT "NO OF COMPARISON PTS (>1)";
460     INPUT M
470     PRINT "LIMITS OF RANGE OF COMPARISON A1,B1 :";
480     INPUT A1,B1
490     H=(B1-A1)/(M-1)
500     PRINT "X VALUE:","FUNCTION:","APPROXN:","ERROR:"
510     FOR I=1 TO M
520     X1=A1+(I-1)*H
530     D=FNF(X1)
540     D1=0
550     FOR J=1 TO P
560     D1=D1+C(J,1)*FNG(X1,J)
570     NEXT J
580     FOR J=P+1 TO N
590     IF Z(J-P) > X1 THEN 610
600     D1=D1+C(J,1)*(X1-Z(J-P))^(P-1)
610     NEXT J
620     E=D-D1
630     PRINT X1,D,D1,E
640     NEXT I
650     DEF FNF(X)=X^3
660     DEF FNG(X,J)=X^(J-1)
670     END

Ready
```

## Sample run 1

```
RUN
SPCOLL

NUMBER OF PARAMETERS? 4
DEGREE OF SPLINE PLUS 1? 3
 1 KNOTS OF SPLINE
? 0
COLLOCATION POINTS X(1),...,X(N)
? -.8
? -.3
? .3
? .8
PARAMETERS C(1),...,C(N)
-.298023E-07
-.24
-1.1
 2.2

NO OF COMPARISON PTS (>1)? 11
LIMITS OF RANGE OF COMPARISON A1,B1 :? -1,1
X VALUE:      FUNCTION:      APPROXN:        ERROR:
-1            -1             -.86            -.14
-.8           -.512          -.512            0
-.6           -.216          -.252           .036
-.4           -.064          -.08            .016
-.2           -.008          .399996E-02     -.012
 0             0             -.298023E-07    .298023E-07
 .2            .008          -.400001E-02    .012
 .4            .064          .08             -.016
 .6            .216          .252            -.360001E-01
 .8            .512          .512             0
 1             1             .86             .14
Ready
```

## Sample run 2

```
OLD SPCOLL
Ready

650      DEF FNF(X)=EXP(X)
RUN
SPCOLL

NUMBER OF PARAMETERS? 6
DEGREE OF SPLINE PLUS 1? 4
 2 KNOTS OF SPLINE
? .5
? 1
COLLOCATION POINTS X(1),...,X(N)
? .125
? .375
? .625
? .875
? 1.125
? 1.375
PARAMETERS C(1),...,C(N)
 1.00055
 .994569
 .504619
 .199326
 .155249
 .204071

NO OF COMPARISON PTS (>1)? 13
LIMITS OF RANGE OF COMPARISON A1,B1 :? 0,1.5
X VALUE:      FUNCTION:      APPROXN:        ERROR:
 0            1              1.00055         -.553608E-03
 .125         1.13315        1.13315         -.238419E-06
 .25          1.28403        1.28385         .17643E-03
 .375         1.45499        1.45499         .119209E-05
 .5           1.64872        1.64891         -.187159E-03
 .625         1.86825        1.86824         .333786E-05
 .75          2.117          2.11684         .155687E-03
 .875         2.39888        2.39887         .524521E-05
 1            2.71828        2.71847         -.191688E-03
 1.125        3.08022        3.08021         .810623E-05
 1.25         3.49034        3.49022         .118732E-03
 1.375        3.95508        3.95507         .104904E-04
 1.5          4.48169        4.48128         .407219E-03
Ready
```

*Program notes*

(1) This Program has been formed by adding new instructions 10, 60, 80, 150–200, 290, 320–350, 550, 580–610 (subject to renumbering) to Program 7.1.

(2) In the Sample runs, a quadratic spline with one knot ($x = 0$) is fitted to $x^3$ on $[-1, 1]$ and a cubic spline with two knots ($x = .5$, 1) is fitted to $e^x$ on $[0, 1.5]$.

### 7.6.2 *Spline collocation with end conditions*

Of course it is not obvious at which points to collocate a spline when using the straightforward scheme of Section 7.6.1. However, there is a natural scheme for collocation, which we shall briefly describe for a cubic spline, and which provides a standard procedure and is very efficient.

Suppose that knots $z_k (k = 1, \ldots, n - 4)$ and end points $z_0 = a$ and $z_{n-3} = b$ are equally spaced with spacing $h$ (so that $z_{k+1} = z_k + h$ for all $k$), and again define $S_k(x)$ by (7.48) as the piece of the spline $S(x)$ in $[z_k, z_{k+1}]$. Then we may form $f*(x)$ by *collocating at the knots and end points* $z_0, \ldots, z_{n-3}$, and also requiring $S'(x)$ to match *given gradients at the end points* $x = a, b$ namely $S'(z_0) = y_0'$, $S'(z_{n-3}) = y_{n-3}'$. A total of $n$ criteria are then specified for determining $n$ parameters $c_1, \ldots, c_n$, and so the spline is uniquely defined. Here $y_k'$ denotes the gradient of $y$ at $x = z_k$.

Briefly, it can be shown that

$$S_k(x) = \frac{1}{6h}\{M_k(z_{k+1} - x)^3 + M_{k+1}(x - z_k)^3$$
$$+ (z_{k+1} - x)(6y_k - M_k h^2) + (x - z_k)(6y_{k+1} - M_{k+1}h^2)\} \quad (7.53)$$

where $y_k = y(z_k)$ are the given data values at the knots and $M_k = S''(z_k)$ $(k = 0, 1, \ldots, n - 3)$.

It is left as an exercise (Problem 15) for the reader to verify that

$$S_k(z_k) = S_{k-1}(z_k), \text{ and } S_k''(z_k) = S_{k-1}''(z_k) = M_k$$
$$(k = 1, \ldots, n - 4) \quad (7.54)$$

Thus (7.53) defines the required spline, provided also that

$$S_k'(z_k) = S_{k-1}'(z_k) \quad (k = 1, \ldots, n - 4) \quad (7.55)$$

The condition (7.55) applied to (7.53) gives the set of equations

$$\left.\begin{array}{r} 2M_0 + M_1 = \alpha_0 \\ M_{k-1} + 4M_k + M_{k+1} = \alpha_k \quad (k = 1, \ldots, n - 4) \\ M_{n-4} + 2M_{n-3} = \alpha_{n-3} \end{array}\right\} \quad (7.56)$$

where

$$\alpha_0 = \frac{6}{h^2}(y_1 - y_0 - hy_0'), \quad \alpha_{n-3} = \frac{6}{h^2}(y_{n-4} - y_{n-3} + hy_{n-3}'),$$

$$\alpha_k = \frac{6}{h^2}(y_{k-1} - 2y_k + y_{k+1}) \quad (k = 1, \ldots, n-4)$$

These are $n-2$ equations for $M_0, \ldots, M_{n-3}$, and, once they have been solved, the spline is given explicitly by (7.53).

Since the system (7.56) has a tridiagonal matrix, it can be solved very efficiently by the special formula described by G. D. Smith (Reference 6, pp. 20–22). Alternatively, since the matrix is diagonally dominant, we can use the Gauss–Seidel or Jacobi iteration to solve (7.56). This is quite convenient and efficient if advantage is taken of the sparseness of the matrix. Indeed Program 6.2 can be used in this case, although it requires slight modifications, since the first and last equations do not have the same diagonal entries and have to be defined separately. The reader is left with the task (Problem 16) of implementing the algorithm.

### 7.6.3  Least squares spline fitting

Finally, especially if we wish to use the r.m.s. error as our norm, a spline of form (7.50) may be fitted by the least squares method to given data $(x_i, y_i)$ $(i = 1, \ldots, m)$, based on Program 7.4. The required modifications are very similar (though fewer in number) to those required in Section 7.6.1 for converting Program 7.1 into Program 7.5 and the details are given below.

**Program 7.6** SPLESQ: Least squares spline fit

```
LIST
SPLESQ

10      REM- SPLESQ: SPECIALISATION OF "LEASQU".
20      REM- LEASQU: FITS C(1)G(X,1)+...+C(N)G(X,N) TO Y VALUES AT
30      REM- X(1),...,X(M) BY LEAST SQUARES, USING BASIC MATRIX ROUTINES
40      REM- F*(X)=SPLINE OF DEG P-1 WITH KNOTS Z(1),...,Z(N-P).
50      DIM X(50),Y(50,1),B(50,10),B1(10,50),A(10,10),C(10,1),U(10,1)
60      DIM Z(10)
70      PRINT "NO OF DATA PTS";
80      INPUT M
90      PRINT "X,Y VALUES:"
100     REM- REDIMENSION Y
110     MAT Y=ZER(M,1)
120     FOR I=1 TO M
130     INPUT X(I),Y(I,1)
140     NEXT I
150     PRINT "NO OF APPROXN PARAMETERS";
160     INPUT N
170     PRINT "DEGREE OF SPLINE PLUS 1";
180     INPUT P
190     PRINT N-P;"KNOTS OF SPLINE"
200     FOR I=1 TO N-P
210     INPUT Z(I)
```

```
220    NEXT I
230    REM- REDIMENSION B,B1,A,C,U
240    MAT B=ZER(M,N)
250    MAT B1=ZER(N,M)
260    MAT A=ZER(N,N)
270    MAT C=ZER(N,1)
280    MAT U=ZER(N,1)
290    REM- FORMS MATRIX B OF OVERDETERMINED EQNS BC=Y
300    REM- G(X,J) IS DEFINED BY FUNCTION FNG(X,J)
310    FOR I=1 TO M
320    FOR J=1 TO P
330    B(I,J)=FNG(X(I),J)
340    NEXT J
350    FOR J=P+1 TO N
360    IF Z(J-P) > X(I) THEN 380
370    B(I,J)=(X(I)-Z(J-P))^(P-1)
380    NEXT J
390    NEXT I
400    MAT B1=TRN(B)
410    REM- SOLVE NORMAL EQNS (B'B)C=B'Y=U
420    MAT A=B1*B
430    MAT U=B1*Y
440    MAT A=INV(A)
450    MAT C=A*U
460    PRINT "PARAMETERS C(1),...,C(N)"
470    FOR I=1 TO N
480    PRINT C(I,1)
490    NEXT I
500    PRINT
510    REM- COMPARE APPROXN C(1)G(X,1)+... WITH Y AT DATA PTS
520    PRINT "X :","GIVEN Y :","FITTED Y :","ERROR :"
530    R=0
540    FOR I=1 TO M
550    D=Y(I,1)
560    D1=0
570    FOR J=1 TO N
580    D1=D1+C(J,1)*B(I,J)
590    NEXT J
600    E=D-D1
610    R=E*E+R
620    PRINT X(I),D,D1,E
630    NEXT I
640    S1=SQR(R/M)
650    PRINT "R.M.S. ERROR :";S1
660    IF M=N THEN 690
670    S2=SQR(R/(M-N))
680    PRINT "ESTIMATED STANDARD DEVIATION :";S2
690    DEF FNG(X,J)=X^(J-1)
700    END

Ready
```

## Sample run 1

```
RUN
SPLESQ

NO OF DATA PTS? 6
X,Y VALUES:
? 0,0
? .2,.192
? .4,.336
? .6,.384
? .8,.288
? 1,0
NO OF APPROXN PARAMETERS? 4
DEGREE OF SPLINE PLUS 1? 3
 1 KNOTS OF SPLINE
? .5
PARAMETERS C(1),...,C(N)
-.001077
 1.13747
-.770644
-1.45871
```

```
X :             GIVEN Y :       FITTED Y :      ERROR :
 0               0              -.001077         .001077
 .2              .192            .195592        -.359215E-02
 .4              .336            .33061          .539017E-02
 .6              .384            .389389        -.538892E-02
 .8              .288            .284407         .359297E-02
 1               0               .107718E-02    -.107718E-02
R.M.S. ERROR : .379094E-02
ESTIMATED STANDARD DEVIATION :  .656611E-02
Ready
```

## Sample run 2

```
RUN
SPLESQ

NO OF DATA PTS? 5
X,Y VALUES:
? .1,.01
? .2,.04
? .3,.09
? .4,.16
? .5,.25
NO OF APPROXN PARAMETERS? 4
DEGREE OF SPLINE PLUS 1? 2
  2 KNOTS OF SPLINE
? .2
? .4
PARAMETERS C(1),...,C(N)
-.166668E-01
 .266673
 .333327
 .333334

X :             GIVEN Y :       FITTED Y :      ERROR :
 .1              .01             .100005E-01    -.528991E-06
 .2              .04             .366678E-01     .333219E-02
 .3              .09             .966678E-01    -.666779E-02
 .4              .16             .156668         .333224E-02
 .5              .25             .250001        -.113249E-05
R.M.S. ERROR : .365149E-02
ESTIMATED STANDARD DEVIATION :  .816497E-02
Ready
```

## Program notes

(1) This Program has been formed by adding new instructions 10, 40, 60, 170–220, 320, 350–380 to Program 7.4 (after suitable re-numbering).

(2) In the Sample runs, a quadratic spline with one knot is fitted to data of $y = x(1 - x^2)$, and a linear spline with 2 knots is fitted to data of $y = x^2$.

### 7.6.4 *Spline basis functions*

The 'elementary' basis functions (7.52), which have been used in our main spline programs 7.5 and 7.6 are far from ideal, and indeed ill-conditioned linear systems are produced unless $n$ is fairly small. Alternative basis functions are therefore normally used, especially 'B-splines'. These lead to well-conditioned linear systems and also to very efficient least squares algorithms, on account of their 'near-orthogonality'. The reader is referred to the text of de Boor (Reference

7) for further information, since B-splines are beyond the scope of this book.

## 7.7 References

1. Mason J.C. *BASIC Numerical Mathematics*, Butterworths (1983).
2. Watson, G.A. *Approximation Theory and Numerical Methods*, John Wiley, Chichester (1980).
3. Snyder, M.A. *Chebyshev Methods in Numerical Approximation*, Prentice-Hall, New Jersey (1966).
4. Cheney, E.W. *Introduction to Approximation Theory*, McGraw-Hill, New York (1966).
5. Forsythe, G.E. *Generation and use of orthogonal polynomials for data-fitting with a digital computer*, SIAM Journal, Vol. 5, pp. 74–88 (1957).
6. Smith, G.D. *Numerical Solution of Partial Differential Equations*, Oxford University Press (1965).
7. de Boor, C. *A Practical Guide to Splines*, Springer Verlag, Berlin (1978).

## PROBLEMS

**(7.1)** Show that the best approximation of form $y = f^*(x) = c_1$ (a straight line parallel to the axis) to the $m$ data below ($m > 2$) is given by (a) $c_1 = 5$, (b) $c_1 = 10/m$, (c) $c_1 = 0$
in the respective cases of the 3 norms: (a) maximum absolute error, (b) r.m.s. error, (c) mean error.
Assuming that the first data point is wild, which of these norms gives the most relevant fit?

| $x$ | 1 | 2 | 3 | 4 | ... | $m$ |
|---|---|---|---|---|---|---|
| $y$ | 10 | 0 | 0 | 0 | ... | 0 |

Hint: Try any other $c_1$ for (a), (c). Use Section 7.5 for (b).

**(7.2)** By using the formula $\cos n\theta = \text{Re}(\cos n\theta + i \sin n\theta) = \text{Re}(\cos \theta + i \sin \theta)^n$, show that $\cos n\theta$ is a polynomial of degree $n$ in $\cos \theta$ with leading coefficient $2^{n-1}$.

What does this tell us about the Chebyshev polynomial $T_n(x)$?

**(7.3)** Prove the Alternating Property (Prop. 1, Section 7.3.2) by showing that $\cos n\theta$ has a similar property.

Prove the minimax property of the Chebyshev polynomials (Prop. 2, Section 7.3.2) by the following steps:

(i) Assume a monic polynomial $\phi_n$ of degree $n$ exists with a smaller maximum magnitude on $[-1, 1]$ than $\psi_n \equiv 2^{1-n} T_n(x)$.

(ii) Prove that $\phi_n - \psi_n$ is a polynomial of degree $n-1$ which is alternatively positive and negative at the $n + 1$ extreme points $x = \cos(n-i)\pi/n$ ($i = 0, 1, \ldots, n$) of $\psi_n$.

(iii) Deduce that $\phi_n - \psi_n$ has $n$ zeros in $[-1, 1]$ and hence that assumption (i) is invalid.

**(7.4)** Find the best approximation in the Chebyshev norm to

(i) $f(x) = 1 + x + x^2$ by a straight line on $[0, 1]$,

(ii) $f(x) = x^3$ by a quadratic on $[-1, 1]$.

(7.5) Determine the 3 zeros of $T_3^*(x)$, and fit a quadratic to $y = \sin \pi x$ on $[0, 1]$ by collocation at these points using Algorithm 7.1.

(7.6) It is claimed that the polynomial of degree $n - 1$, which collocates $f(x) = (1 + 25x^2)^{-1}$ at the $n$ equally spaced points $x = x_i = (i - 1)/(n - 1)$ diverges from $f(x)$ on $[0, 1]$ as $n \to \infty$. Test this claim using Program 7.1. (You may use values of $n$ above 10, if you modify the DIM statement.)

(7.7) Use Algorithm 7.2 instead of Algorithm 7.1 in Problem 5, and, by rewriting the resulting approximation $c_1 T_0^*(x) + \ldots + c_3 T_2^*(x)$ in the form $a_1 + a_2 x + a_3 x^2$, verify that it is essentially identical to that produced by Algorithm 7.1.

(7.8) A simple tensile test on a carbon steel cylinder gives the following readings of strain ($\varepsilon$) against stress ($\sigma$). Determine the modulus of elasticity of the cylinder by linear least squares (Algorithm 7.3).

| $\sigma$ (MPa) | 100 | 200 | 300 | 400 | 500 | 600 | 700 |
|---|---|---|---|---|---|---|---|
| $\varepsilon$ (%) | 0.31 | 0.92 | 1.40 | 1.49 | 1.95 | 2.26 | 2.80 |

(7.9) If a grain of propellant is dropped onto a hot surface at temperature $T°$ (K) it ignites after time $t$ (s). A typical set of observed data is given below for which the physical model is

$$t = A \exp[B/(R_0 T)]$$

where $A$ and $B$ are constants and $R_0 = 8314$ J/KgK. By rectifying the data and using a linear least squares fit, determine $\log_e A$ and $B$.

| $T°$ (K) | 483 | 488 | 493 | 498 | 503 | 508 | 513 | 518 | 523 |
|---|---|---|---|---|---|---|---|---|---|
| $t$ (s) | 111 | 79 | 50 | 29 | 16 | 10 | 7 | 5.3 | 3.0 |

(7.10) In the table below are 13 data in non-dimensional units relating the velocity $y$ at a point of a fluid to its distance $x$ from a flat plate.

These data were obtained from a solution of the Blasius equation valid in a boundary layer.

Use Program 7.4 to obtain a least squares fit to the decaying function $1 - y$ by

$$f^*(x) = c_1 e^{-x} + c_2 e^{-2x} + \ldots + c_n e^{-nx} \text{ for } n = 1, 2, 4.$$

| $x$ | 0 | 0.6 | 1.2 | 1.8 | 2.4 | 3.0 | 3.6 |
|---|---|---|---|---|---|---|---|
| $y$ | 0 | .1989 | .3938 | .5748 | .7290 | .8460 | .9233 |
| $x$ | 4.2 | 4.8 | 5.4 | 6.0 | 6.6 | 7.2 | |
| $y$ | .9670 | .9878 | .9962 | .9990 | .9998 | 1.0000. | |

(7.11) The set of orthogonal polynomials $\{g_j(x)\}$ which satisfy (7.40) can be defined efficiently for any set of data points $x_i (i = 1, \ldots, m)$ by the relations

$$g_1(x) = 1, \quad g_{j+1}(x) = (x - B_j)g_j(x) + C_j g_{j-1}(x) \quad (j = 1, 2, \ldots) \quad (*)$$

where $\qquad S_j = \sum_{k=1}^{m} [g_j(x_k)]^2$, $B_j = \sum_{k=1}^{m} x_k[g_j(x_k)]^2/S_j$;

and $\qquad\qquad C_1 = 0$, $C_j = S_j/S_{j-1}$ for $j \geqslant 2$.

Verify that these relations are valid by induction (i.e. by showing that if, for any $k$, $g_1(x)$, ..., $g_k(x)$ are orthogonal, then $g_1(x)$, ..., $g_{k+1}(x)$ are orthogonal as a consequence of (*)).

**(7.12)** Write a program which obtains a best least squares fit to a set of data by using orthogonal polynomials, based on either the Gram–Schmidt method of Section 7.5.3 or the more efficient method of Problem 11.

**(7.13)** Find a cubic spline approximation to $\log_e (1 + x)$ on [0, 1], with one knot at 0.5, by collocation at 5 equally spaced points $x_1, ..., x_5$ $(0 < x_1 < ... < x_5 < 1)$ and compare your results at 21 points.

By adjusting the values of $x_1, ..., x_5$ (i.e. no longer equally spaced) try to obtain an approximation for which the extreme values of the error are all nearly equal. (Program 7.5)

**(7.14)** Use the least squares method (Program 7.6) to find a spline of degree 1 with 2 knots (i.e. a set of 3 straight lines joined together) which is the best fit to the data of Problem 10.

Adjust the positions of the knots to give the most satisfactory results.

**(7.15)** Verify that the continuity relations (7.54) hold for the piecewise cubic given by (7.53).

Use the algorithm of Section 7.6.2 to determine a cubic spline $S(x)$ with one knot at $x = 0$ which collocates $y = \sin x$ at $x = \pm\pi/2$, 0 and matches its gradient at $x = \pm\pi/2$.

*Check:* $S(x)$ coincides with the cubic polynomial $y = -4x^3/\pi^3 + 3x/\pi$.

**(7.16)** Write a program to implement the algorithm of Section 7.6.2 and test it on Problem 15.

# Index